Crossroads between Innate and Adaptive Immunity

ADVANCES IN EXPERIMENTAL MEDICINE AND BIOLOGY

Recent Volumes in this Series

Volume 582
HOT TOPICS IN INFECTION AND IMMUNITY IN CHILDREN III
Edited by Andrew J. Pollard and Adam Finn

Volume 583
TAURINE 6
Edited by Simo S. Oja and Pirjo Saransaari

Volume 584
LYMPHOCYTE SIGNAL TRANSDUCTION
Edited by Constantine Tsoukas

Volume 585
TISSUE ENGINEERING
Edited by John P. Fisher

Volume 586
CURRENT TOPICS IN COMPLEMENT
Edited by John D. Lambris

Volume 587
NEW TRENDS IN CANCER FOR THE 21st CENTURY
Edited by Antonio Llombart-Bosch, Jose López-Guerrero and
Vincenzo Felipe

Volume 588
HYPOXIA AND EXERCISE
Edited by Robert C. Roach, Peter D. Wagner, and
Peter H. Hackett

Volume 589
NEURAL CREST INDUCTION AND DIFFERENTIATION
Edited by Jean-Pierre Saint-Jeannet

Volume 590
CROSSROADS BETWEEN INNATE AND ADAPTIVE IMMUNITY
Edited by Peter D. Katsikis, Bali Pulendran and
Stephen P. Schoenberger

Peter D. Katsikis
Bali Pulendran
Stephen P. Schoenberger

Editors

Crossroads between Innate and Adaptive Immunity

With 39 Illustrations

 Springer

Peter D. Katsikis
Drexel University College of
    Medicine
Philadelphia, Pennsylvania
USA
peter.katsikis@drexelmed.edu

Stephen P. Schoenberger
La Jolla Institute for Allergy and
Immunology
La Jolla, California
USA
sps@liai.org

Bali Pulendran
Emory University
Emory Vaccine Center
Atlanta, Georgia
USA
bpulend@rmy.emory.edu

Library of Congress Control Number: 2006931119
Printed on acid-free paper.

ISBN-10: 0-387-34813-1   e-ISBN-10: 0-387-34814-X
ISBN-13: 978-0387-34813-1   e-ISBN-13: 978-0387-34814-8

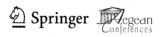 Springer

Proceedings of the "First Crossroads between Innate and Adaptive Immunity" Conference,
held in Rhodes, Greece, October 9-14, 2005.

9 8 7 6 5 4 3 2 1

springer.com

Preface

This compilation presents minireviews derived from work presented at the Aegean Conference: "First Crossroads between Innate and Adaptive Immunity," which occurred October 9–14, 2005 at the Hilton Conference Center on the island of Rhodes, Greece. The conference included sessions dedicated to host recognition of and response to pathogens, innate immune networks, antigen presentation, and adaptive immune responses, each headlined by a leading scientist at the forefront of that field. The intimate networking and interaction of dendritic cells, T cells, B cells, NK cells, NK-T cells, and complement during the host response to pathogens and tumors are only now starting to be elucidated. The respective fields that focus on these immune cells and molecules tend to exist in parallel worlds, with minimum exchange of information and ideas. The goal of this conference was to initiate crosstalk between these immunological fields, and to promote and facilitate discussion on the interactions between the innate immune response and the adaptive immune response, and ultimately breed collaboration between these areas of study. The "First Crossroads between Innate and Adaptive Immunity" Aegean Conference succeeded in bringing together and connecting scientists and experts from around the world to address critical areas of innate and adaptive immunity — something necessary for the development of more efficient scientific exchange and crosspollination between these fields.

Acknowledgments

We would like to thank all of the researchers and presenters who attended the conference and made the conference as enlightening and enjoyable as it was. We especially would like to thank those who contributed chapters to this book and the people at Springer for giving us the opportunity to publish these proceedings in their Advanced Experimental Medicine and Biology book series. Special thanks to Douglas Dolfi for his help in editing and formatting chapters for the book. We would also like to thank the people of Aegean Conferences, and especially Dimitrios Lambris, who took care of everyone and everything, making the entire conference go seamlessly. Finally, we would like to acknowledge the generous support from 3M, Drexel University, eBioscience, Gemini Science Inc., Genentech, the Institute for Molecular Medicine and Infectious Disease at Drexel University College of Medicine, Wyeth, and the Aegean Conferences, without whose support this meeting would not have been possible. Thank you all, and we hope to see you again in 2007.

Peter D. Katsikis. MD, PhD
Bali Pulendran, PhD
Stephen Schoenberger, PhD

Contents

List of Contributors xvii

1. **Signal Transduction in DC Differentiation:**
   Winged Messengers and Achilles' Heel
   Inna Lindner, Pedro J. Cejas, Louise M. Carlson,
   Julie Torruellas, Gregory V. Plano, and Kelvin P. Lee
   1. Introduction... 1
   2. Dendritic Cell Functional Diversity 1
   3. DC Progenitors 6
   4. Winged Messengers — Signaling Pathways that Drive
      DC Differentiation 6
      4.1. Extracellular Stimuli.. 6
      4.2. Intracellular Signal Transduction 7
   5. Achilles' Heel — Subversion of DC Differentiation
      by Pathogen-Mediated Disruption of Signal
      Transduction Pathways 15
      5.1. Yersinia and the Disruption of Intracellular Signaling Pathways....... 16
      5.2. Yersinia and DC Differentiation.. 16
   6. Concluding Remarks... 17
   7. References... 17

2. **Shaping Naive and Memory CD8+ T Cell Responses**
   in Pathogen Infections through Antigen Presentation
   Gabrielle T. Belz, Nicholas S. Wilson, Fiona Kupresanin,
   Adele M. Mount, and Christopher M. Smith
   1. Introduction... 31
   2. Dendritic Cells of Spleen and Lymph Nodes............................... 31
   3. Role of Dendritic Cells in Pathogen Responses 33
      3.1. Priming Naive T Cells.. 33
      3.2. Identifying the Main Movers and Shakers in Infection 34
      3.3. Dendritic Cell Subsets in Pathogen Infections.................... 35
      3.4. Amplification of Memory CD8+ T Cells in Secondary Infections..... 38
   4. Conclusions... 39
   5. Acknowledgments.. 39
   6. References... 39

3. **Understanding the Role of Innate Immunity in**
    the Mechanism of Action of the Live Attenuated
    Yellow Fever Vaccine 17D
    Troy D. Querec and Bali Pulendran
    1. Introduction: A Historical Perspective 43
    2. Understanding the Innate Immune Mechanism
        of Action of YF-17D 46
    3. Concluding Remarks 48
    4. Acknowledgments 48
    5. References 48

4. **The Function of Local Lymphoid Tissues in**
    Pulmonary Immune Responses
    Juan Moyron-Quiroz, Javier Rangel-Moreno,
    Damian M. Carragher, and Troy D. Randall
    1. Lymph Node Structure and Development 55
    2. Role of Local Lymphoid Organs in Pulmonary Immunity 57
        2.1. Structure and Function of Nasal-Associated Lymphoid
              Tissue (NALT) 57
        2.2. Pulmonary Immune Responses in the Absence of
              Secondary Lymphoid Organs 58
        2.3. Structure and Function of Bronchus-Associated
              Lymphoid Tissue (BALT) 60
        2.4. Does iBALT Confer Antiinflammatory Properties on
              Local Immune Responses? 63
    3. Conclusions and Future Directions 63
    4. Acknowledgments 64
    5. References 65

5. **The Yin and Yang of Adaptive Immunity in**
    Allogeneic Hematopoietic Cell Transplantation:
    Donor Antigen-Presenting Cells Can Either
    Augment or Inhibit Donor T Cell Alloreactivity
    Jian-Ming Li and Edmund K. Waller
    1. Introduction 70
    2. Materials and Methods 71
        2.1. Mice 71
        2.2. LBRM Tumor Cell Line 72
        2.3. Donor Cell Preparations 72
        2.4. BM CD11b Depletion and Splenic T Cell Purification 73
        2.5. Recipient Mice Conditioning 73
        2.6. BMT and Leukemia Challenge 73
        2.7. Analyses of DC Subsets and DC Precursors in BM
              and Spleen Grafts 73
        2.8. Analyses of Hematopoietic Engraftment of
              Transplant Recipients 74

2.9. Serum Gamma Interferon (IFN-γ) and Tumor
Necrosis Factor-Alpha (TNF-α)
Enzyme-Linked Immunosorbent Assay (ELISA).............................. 74
2.10. Assessments of Survival and GvHD in Transplant Recipients 74
2.11. Statistical Analyses... 74
3. Results.. 75
3.1. MACs Depletion of CD11b⁺ Cells in the BM Graft
Does not Affect Stem Cell Content 75
3.2. Transplanting Manipulated BM Grafts in the
Absence of Added Splenocytes Did not Lead
to Graft Rejection or GvHD 75
3.3. CD11b-Depleted BM Grafts Combined with Low-Dose
Splenocytes or Splenic T Cells Led to Slight Enhancement
of Non-Lethal GvHD in Recipients of Allogeneic BMT.................... 75
3.4. CD11b⁺ Cell-Enriched BM Grafts Combined with
Low-Dose Splenocytes or Splenic T Cells Inhibited
GvHD in Recipients of Allogeneic BMT 76
3.5. No Combinations of Unmanipulated BM and Splenocytes
Produced a GvL Effect without also Causing Lethal GvHD 77
3.6. The Combination of CD11b-Depleted BM and
Low-Dose Donor Splenocytes Led to a Durable
GvL Effect without GvHD 78
3.7. Recipients of CD11b-Depleted BM Grafts Had
Increased Numbers of Donor Spleen-Derived
Memory T Cells in the Blood Post-Transplant 79
3.8. Recipients of CD11b-Depleted Allogeneic BMT Had
Increased Levels of Serum IFN-γ at Day +30 Post-BMT 81
4. Discussion... 81
5. Acknowledgments.. 84
6. References... 84

6. It's Only Innate Immunity But I Like It
Emanuela Marcenaro, Mariella Della Chiesa,
Alessandra Dondero, Bruna Ferranti, and
Alessandro Moretta
1. Introduction... 89
2. The Immunoregulatory Role of NK Cells:
Crosstalk between NK, MDDC, and PDC 90
3. Crosstalk between Innate and Adaptive Immune Responses 92
4. Involvement of Neutrophils in the Regulation
of Adaptive Immune Responses through
Interactions with Other Innate Effector Cells.................................... 93
5. Other Innate Cells Such as Mast Cells or
Eosinophils Are Important in the Early
Phases of Innate Immune Responses 94
6. Concluding Remarks... 97
7. Ackowledgments.. 97
8. References.. 97

7.   **Innate Tumor Immune Surveillance**
     Mark J. Smyth, Jeremy Swann, and
     Yoshihiro Hayakawa
     1.   Introduction..   103
     2.   Type I Interferon   103
     3.   NKG2D...   104
     4.   Cytokines that Act via NKG2D   106
     5.   Acknowledgments..   106
     6.   References...   107

8.   **Regulation of Adaptive Immunity by Cells of the**
     Innate Immune System: Bone Marrow Natural
     Killer Cells Inhibit T Cell Proliferation
     Prachi P. Trivedi, Taba K. Amouzegar, Paul C. Roberts,
     Norbert A. Wolf, and Robert H. Swanborg
     1.   Introduction..   113
     2.   Regulatory Function of NK Cells..   114
     3.   NK Cells Inhibit by a Non-Cytotoxic Mechanism.............................   115
     4.   NK Cells Inhibit Cell Cycle Progression   116
     5.   Bone Marrow-Derived NK Cells Have
          Unique Function..   117
     6.   Role of NK Cells in Immune Homeostasis   118
     7.   Acknowledgments..   119
     8.   References...   119

9.   **Induction and Maintenance of CD8+ T Cells**
     Specific for Persistent Viruses
     Ester M.M. van Leeuwen, Ineke J.M. ten Berge,
     and René A.W. van Lier
     1.   Persistent Viruses Are Prevalent in Human and Mice.........................   122
     2.   General Effects of Persistent Viruses on
          the Host Immune System   123
     3.   Generation of CD8+ Memory T Cells..   124
     4.   Function of Memory CD8+ T Cells Specific
          for Persistent Viruses   126
     5.   Phenotype of Memory CD8+ T Cells Specific
          for Persistent Viruses   127
     6.   Maintenance of Memory CD8+ T Cells Specific
          for Persistent Viruses   129
     7.   Regulation of IL-7Rα Expression by the
          Presence of Antigen   130
     8.   Concluding Remarks..   131
     9.   References...   131

10. Germinal Center-Derived B Cell Memory
Craig P. Chappell and Joshy Jacob
1. Introduction... 139
2. Materials and Methods.. 140
   2.1. Mice and Immunizations 140
   2.2. β-Galactosidase Detection, Antibodies,
        and Flow Cytometry... 140
   2.3. ELISPOT Assay 141
   2.4. Cell Sorting and Adoptive Transfers 141
   2.5. PCR and DNA Sequencing.. 141
   2.6. Statistics 141
3. Results.. 141
   3.1. Generation of Germinal Center-Cre Transgenic Mice.................... 141
   3.2. Splenic β-gal Expression Is Induced upon Immunization 142
   3.3. β-Galactosidase Expression Does not Mark All GC B Cells.......... 143
   3.4. β-gal⁺ GC B Cells Contain Mutated λ_1 V Regions....................... 144
   3.5. Hypermutated β-gal⁺ Memory B Cells Transfer
        Ag Recall Responses 145
4. Conclusion 147
5. References.. 147

11. CD28 and CD27 Costimulation of CD8+ T Cells: A Story of Survival
Douglas V. Dolfi and Peter D. Katsikis
1. Introduction... 149
2. Classical and Alternative Costimulation 149
3. T Cell Development... 151
4. Antigen-Specific T Cell Responses... 155
5. Memory, Antigenic Rechallenge, and
   Secondary Responses.. 160
6. Summary and Conclusions.. 163
7. References.. 163

12. CD38: An Ecto-Enzyme at the Crossroads of Innate and Adaptive Immune Responses
Santiago Partida-Sánchez, Laura Rivero-Nava, Guixiu Shi, and Frances E. Lund
1. Introduction... 171
2. CD38 Regulates Innate and Adaptive
   Immune Responses 172
   2.1. CD38 Regulates Neutrophil Migration and Lung
        Inflammatory Responses 172
   2.2. CD38 Regulates Dendritic Cell Trafficking
        in Vitro and in Vivo.. 173
   2.3. CD38 Regulates T Cell-Dependent Immune Responses 174
3. CD38 Modulates Chemokine Receptor Signaling
   by Producing Calcium Mobilizing Metabolites 176

4. Conclusions and Future Directions 178
    4.1. Unresolved Questions 178
    4.2. Model 179
5. Acknowledgments 180
6. References 180

13. **Vascular Leukocytes: A Population with
    Angiogenic and Immunossuppressive Properties
    Highly Represented in Ovarian Cancer**
    *George Coukos, Jose R. Conejo-Garcia,
    Ron Buckanovich, and Fabian Benencia*
    1. Physiological Angiogenesis vs. Pathological Angiogenesis 185
    2. Endothelial Progenitors and Neoangiogenesis 186
    3. Hematopoietic Cells Participate in Neoangiogenesis 186
    4. Antigen-Presenting Cells as Endothelial Cells 186
    5. Tumor Angiogenesis 187
    6. Vascular Leukocytes 187
    7. Vascular Leukocytes and Antitumor Immune Response 190
    8. Final Remarks 191
    9. Acknowledments 191
    10. References 191

14. **CD4+ T Cells Cooperate with Macrophages
    for Specific Elimination of MHC Class
    II-Negative Cancer Cells**
    Alexandre Corthay
    1. Introduction 195
    2. CD4+ T Cells Help CD8+ T Cells to Kill Tumor Cells 196
    3. CD4+ T Cells Can Reject Tumors in the Absence
        of CD8+ T Cells 196
    4. Cancer Immunotherapy by Adoptive Transfer of
        Tumor-Specific CD4+ T Cells 196
    5. CD4+ T Cells in Cancer Immunosurveillance 197
        5.1. Injection of Tumor Cells in Matrigel 197
        5.2. Naive Tumor-Specific CD4+ T Cells Become Activated in
              Draining Lymph Nodes (LN), Migrate to the Incipient
              Tumor Site and Secrete Cytokines 199
        5.3. Massive Recruitment of Host Macrophages toward the
              Injected Myeloma Cells 200
        5.4. Tumor-Specific CD4+ T Cells Activate
              Matrigel-Infiltrating Macrophages 201
        5.5. IFNγ Is Critical for T Cell-Mediated Macrophage
              Activation and Tumor Rejection 202
        5.6. T Cell-Activated Macrophages Suppress Tumor
              Cell Growth 203
    6. Conclusions 204
    7. Acknowledgments 205
    8. References 205

15. **Receptors and Pathways in Innate Antifungal Immunity:**
    The Implication for Tolerance and Immunity to Fungi
    Teresa Zelante, Claudia Montagnoli, Silvia Bozza,
    Roberta Gaziano, Silvia Bellocchio, Pierluigi Bonifazi,
    Silvia Moretti, Francesca Fallarino,
    Paolo Puccetti, and Luigina Romani
    1. Introduction 209
    2. What and Which Are Opportunistic
       Fungal Pathogens? 210
    3. The Immune Response to Fungi: From Microbe
       Sensing to Host Defencing 210
    4. Sensing Fungi 211
    5. Tuning the Adaptive Immune Responses:
       the Instructive Role of DCs 214
    6. DCs as Tolerance Mediators via Tryptophan
       Catabolism 215
    7. Dampening Inflammation and Allergy to
       Fungi through Treg 216
    8. Looking Forward 218
    9. Acknowledgments 219
    10. References 219

Author Index 223

Subject Index 225

List of Contributors

Taba K. Amouzegar
Department of Immunology and Microbiology
Wayne State University
School of Medicine
Detroit, Michigan, USA

Silvia Bellocchio
Department of Experimental Medicine
and Biochemical Sciences
University of Perugia
Perugia, Italy

Gabrielle T. Belz
Division of Immunology
The Walter and Eliza Hall Institute
of Medical Research
Melbourne, Australia

Fabian Benencia
Abramson Family Cancer Research
Institute
University of Pennsylvania
Philadelphia, Pennsylvania, USA

Pierluigi Bonifazi
Department of Experimental Medicine
and Biochemical Sciences
University of Perugia
Perugia, Italy

Silvia Bozza
Department of Experimental Medicine
and Biochemical Sciences
University of Perugia
Perugia, Italy

Ron Buchanovich
Center for Research on Reproduction
and Women's Health
University of Pennsylvania
Philadelphia, Pennsylvania, USA

Louise M. Carlson
University of Miami Miller School
of Medicine
Miami, Florida, USA

Damian M. Caragher
Trudeau Institute
Saranac Lake, New York, USA

Pedro J. Cejas
University of Miami Miller School
of Medicine
Miami, Florida, USA

Craig P. Chappell
Department of Microbiology
and Immunology
Emory University
Atlanta, Georgia, USA

Mariella Della Chiesa
Dipartimento di Medicina Sperimentale
Università degli Studi di Genova
Genova, Italy

Jose R. Conejo-Garcia
Center for Research on Reproduction
and Women's Health
University of Pennsylvania
Philadelphia, Pennsylvania, USA

Alexandre Corthay
Institute of Immunology
University of Oslo and Rikshospitalet-
  Radiumhospitalet Medical Center
Oslo, Norway

George Coukos
Center for Research on Reproduction
  and Women's Health
University of Pennsylvania
Philadelphia, Pennsylvania, USA

Douglas V. Dolfi
Department of Microbiology and
  Immunology
Drexel University College of Medicine
Philadelphia, Pennsylvania, USA

Alessandra Donder
Dipartimento di Medicina Sperimentale
Università degli Studi di Genova
Genova, Italy

Francesca Fallarino
Department of Experimental Medicine
  and Biochemical Sciences
University of Perugia
Perugia, Italy

Bruna Ferranti
Dipartimento di Medicina Sperimentale
Università degli Studi di Genova
Genova, Italy

Roberta Gaziano
Department of Experimental Medicine
  and Biochemical Sciences
University of Perugia
Perugia, Italy

Yoshihiro Hayakawa
Cancer Immunology Program
Peter MacCallum Cancer Centre
East Melbourne, Victoria, Australia

Joshy Jacob
Department of Microbiology and
  Immunology
Emory University
Atlanta, Georgia, USA

Peter D. Katsikis
Department of Microbiology and
  Immunology
Drexel University College of Medicine
Philadelphia, Pennsylvania, USA

Fiona Kupresanin
Division of Immunology
The Walter and Eliza Hall Institute
  of Medical Research
Melbourne, Australia

Kelvin P. Lee
University of Miami Miller School
  of Medicine
Miami, Florida, USA

Jian-Ming Li
Emory University
Atlanta, Georgia, USA

Inna Lindner
University of Miami Miller School
  of Medicine
Miami, Florida, USA

Frances E. Lund
Trudeau Institute
Saranac Lake, New York, USA

Emanuela Marcenaro
Dipartimento di Medicina Sperimentale
Università degli Studi di Genova
Genova, Italy

Claudia Montagnoli
Department of Experimental Medicine
  and Biochemical Sciences
University of Perugia
Perugia, Italy

Alessandro Moretta
Dipartimento di Medicina Sperimentale
Istituto Giannina Gaslini
Centro di Eccellenza per le Ricerche
  Biomediche
Università degli Studi di Genova
Genova, Italy

Silvia Moretti
Department of Experimental Medicine
and Biochemical Sciences
University of Perugia
Perugia, Italy

Adele M. Mount
Division of Immunology
The Walter and Eliza Hall Institute
of Medical Research
Melbourne, Australia

Juan Moyron-Quiroz
Trudeau Institute
Saranac Lake, New York, USA

Santiago Partida-Sánchez
Columbus Children's Research
Institute
Columbus, OH, USA

Gregory V. Plano
University of Miami Miller School
of Medicine
Miami, Florida, USA

Paolo Puccetti
Department of Experimental Medicine
and Biochemical Sciences
University of Perugia
Perugia, Italy

Bali Pulendran
Department of Pathology & Emory
Vaccine Center
Atlanta, Georgia, USA

Troy D. Querec
Department of Pathology & Emory
Vaccine Center
Atlanta, Georgia, USA

Troy Randall
Trudeau Institute
Saranac Lake, New York, USA

Javier Rangel-Moreno
Trudeau Institute
Saranac Lake, New York, USA

Laura Rivero-Nava
Trudeau Institute
Saranac Lake, New York, USA

Paul C. Roberts
Department of Immunology
and Microbiology
Wayne State University School
of Medicine
Detroit, Michigan, USA

Luigina Romani
Department of Experimental Medicine
and Biochemical Sciences
University of Perugia
Perugia, Italy

Guixiu Shi
Trudeau Institute
Saranac Lake, New York, USA

Christopher M. Smith
Department of Pathology
University of Cambridge
Cambridge, UK

Mark J. Smyth
Cancer Immunology Program
Peter MacCallum Cancer Centre
East Melbourne, Victoria, Australia

Robert H. Swanborg
Department of Immunology
and Microbiology
Wayne State University School
of Medicine
Detroit, Michigan, USA

Jeremy Swann
Cancer Immunology Program
Peter MacCallum Cancer Centre
East Melbourne, Victoria, Australia

Ineke J.M. ten Berge
Department of Internal Medicine
Academic Medical Center
Amsterdam, The Netherlands

Julie Torruellas
University of Miami Miller School
    of Medicine
Miami, Florida, USA

Prachi P. Trivedi
Department of Molecular and Cellular
    Biology
Harvard University
Cambridge, Massachusetts, USA

Ester M.M. van Leeuwen
Department of Experimental
    Immunology
Academic Medical Center
Amsterdam, The Netherlands

René A.W. van Lier
Department of Experimental
    Immunology
Academic Medical Center
Amsterdam, The Netherlands

Edmund K. Waller
Emory University
Atlanta, Georgia, USA

Nicholas S. Wilson
CSL Limited
Melbourne, Australia

Norbert A. Wolf
Department of Immunology
    and Microbiology
Wayne State University School
    of Medicine
Detroit, Michigan, USA

Teresa Zelante
Department of Experimental Medicine
    and Biochemical Sciences
University of Perugia
Perugia, Italy

1

SIGNAL TRANSDUCTION IN DC DIFFERENTIATION:

WINGED MESSENGERS AND ACHILLES' HEEL

Inna Lindner, Pedro J. Cejas, Louise M. Carlson, Julie Torruellas, Gregory V. Plano, and Kelvin P. Lee*

1. INTRODUCTION

Dendritic cells (DC) are centrally involved in the initiation and regulation of the adaptive immune response, and different DC can have markedly different (e.g., opposing) function. Acquisition of specific functions is likely to be a result of both nature and nurture, namely differentiation of progenitors into distinct DC subsets as well as the influence of environmental signals. This is not unlike what is seen for T and B cells. This review will focus on the signal transduction pathways that allow an unusually wide range of hematopoietic progenitors to differentiate into DC, the functional characteristics regulated by these pathways, and the ability of pathogens to alter DC function by subverting these pathways during progenitor→DC differentiation.

2. DENDRITIC CELL FUNCTIONAL DIVERSITY

Dendritic cells are professional antigen-presenting cells that play an essential role in the activation of T lymphocytes in response to foreign antigens, in the deletion of the autoreactive T cells in the thymus (central tolerance)[1,2], and in the induction of T cell anergy and tolerance in the periphery (peripheral tolerance)[3,4]. Evidence suggests that generation of DC occurs in two stages: (1) differentiation of a progenitor to the immature/unactivated DC (iDC) (lineage commitment) whose function is to continuously sample the antigenic environ-

*University of Miami Miller School of Medicine, Miami, FL 33136, USA. Address correspondence to: Klee@med.miami.edu

ment, and (2) activation (or maturation) of immature DC by "danger signals" such as microbial products, tissue necrosis, and proinflammatory cytokines[4-7]. In this review, we will use "mature" and "activated" (and conversely "immature" and "unactivated") synonymously when referring to dendritic cells, even though there are subtle semantic differences. As DC undergo maturation, they down-regulate their ability to uptake antigen, upregulate expression of MHC class I and II, express high levels of costimulatory ligands (CD40, CD80, and CD86) and emigrate from the peripheral tissues to secondary lymphoid organs, where they present to and activate antigen-specific T cells (as well as other immune cells)[8]. In addition to inducing T cell proliferation, DC also dictate the qualitative outcome of the T cell response. In the secondary lymphoid organs, the T cell activating signals from DC together with the local cytokine milieu can regulate the development of functionally distinct CD4 T cell subsets. These polarized CD4 cells can secrete a different pattern of cytokines and subsequently regulate cellular (Th1) response vs. humoral (Th2) response[9]. The most characteristic cytokine produced by Th1 cells is IFN-γ, while IL-4, IL-5, IL-10, and IL-13 are generally produced by the Th2 CD4 cells. An important factor in Th1 development is the production of IL-12 by the DC[10], which is positively controlled by the engagement of CD40 by the CD40 ligand, CD154, and can be negatively regulated by Th2 cytokines such as IL-10 and IL-4[11-13].

Thus, one level of DC functional diversity comes from the maturation status of the DC. Broadly, iDC act to maintain tolerance, whereas mature DC have strong immunostimulatory properties (Fig. 1). Mature, activated DC are unique in their capacity to potently prime naive T cells. In mice expressing transgenic T cell receptor, DC induce a primary T cell response to soluble antigen in vitro that is 100- to 300-fold more potent than that induced by any other APC[14,15]. It is becoming increasingly clear that the balance of tolerance vs. activation is even more complex, involving not only the state of DC maturation but also its local microenvironment[16] (Fig. 2). For example, while mature DC very effectively induce T cell activation, immature DC can induce T cell unresponsiveness, delete antigen specific T cells, or generate regulatory T cells (Treg)[17-19].

Although DC maturation typically enhances immunostimulatory capacity, several exogenous factors can significantly influence this. For example, the presence of IL-10 during DC maturation results in DC that generate anergic T cells, characterized by inhibition of proliferation, and decreases both in the expression of CD25 and production of IL-2[20]. In addition to cytokines, DC grown in the presence of other factors such as prostaglandin E2 and corticosteroids are unable to secrete IL-12p70 and skew the Th1/Th2 balance toward Th2[21-23]. In vivo, local immunosuppressive effects characteristic of particular tissues can also induce tolerogenic DC. Tumor-associated DC, for example, have low allostimulatory capacity and have an immature phenotype, possibly diverting immune response towards tolerance[24]. Similarly liver-derived DC have been shown to prolong islet allograft survival — probably through the induction of some kind of regulatory

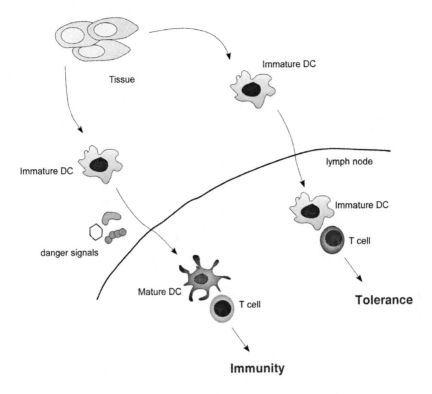

Figure 1. DC functions. During the steady-state conditions (absence of infection or inflammation), peripheral tissue-resident DC capture and process self antigen while remaining in the immature state. When these immature DC travel to the lymph nodes they present antigen to the naive T cells in the absence of costimulatory molecules and subsequently induce T cell tolerance. Microbial infection, inflammation, and tissue damage induce DC maturation and migration to the lymph nodes. The mature DC express both antigen presentation and costimulatory molecules, which allow priming and activation of naive T cells in the lymph node.

T cell[25]. In addition to the local microenvironment, the functional specialization of DC also stems from their differential expression of pattern-recognition receptors such as Toll-like receptors (TLR) and subsequently their selective recognition of various microbial products.

Although the process of DC activation and the conditions under which maturation occurs appear to be important in determining the fate of the DC and the subsequent T cell response, recent studies suggest that the modulation of transcription factor expression that initiate DC differentiation from precursors also modifies the function of DC[26,27]. Two recent reports demonstrated that inhibiting NFκB during DC development results in a suppressed primary immune response[26,27] and that in a murine model the NFκB-deficient DC induce generation of regulatory T cells[26]. NFκB inhibition appears to carry biological signifi-

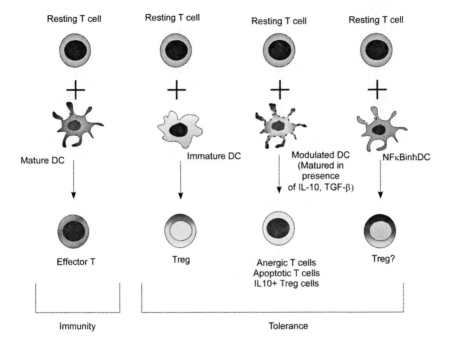

Figure 2. Activation of T cells by functionally different DC populations results in the development of distinct T cell responses. Mature DC induce the development of immunostimulatory effector T cells while immature DC generate the development of regulatory T cells (Treg). The presence of certain factors such as IL-10 and TGF-β during DC maturation stage modulates DC development and results in generation of tolerogenic T cells. Inhibition of the NFκB pathway (NFκBinhDC) during DC differentiation from precursors has also been shown to result in generation of Treg in certain systems.

cance in cases of tumorogenesis where tumor-secreted factors inhibit DC development by suppressing NFκB[28]. These findings suggest that events which occur prior to DC activation may mold the subsequent DC response.

Two models have been proposed to explain how functionally distinct DC are generated (Figure 3)[29]. One model proposes that the functional plasticity of DC arises from the different activation states of a common immature DC. One of the important factors regulating the characteristics of the mature DC and the outcome of the DC–T cell interaction proposed by this model is the combination of signals engaged by the DC during its activation in the local milieu[30]. The second model proposes that there are signals acting very early during the differentiation of hematopoietic cells that allow for the divergence of DC precursors that later develop into distinct DC subsets[31-33]. These developmentally distinct DC are committed to a particular function that may be unique for that subset. This model is supported by a recent finding that early upregulation of lipid-activated

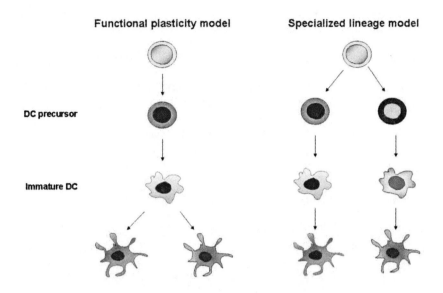

Figure 3. Generation of functionally distinct DC-2 models.

transcription factor peroxisome proliferative activated receptor-γ (PPARγ) during DC differentiation from monocytes specifies a DC subset that stimulates invariant natural killer cells (iNKT)[34], as well as some other findings implicating the expression of various transcription factors and signaling molecules (e.g., PU.1, Notch-1) in development of unique DC populations[35,36]. Interestingly, in tumor-bearing hosts there is a DC differentiation block that results in accumulation of immature myeloid cells (which include DC progenitors), which themselves are immunosuppressive[37] — suggesting that even DC precursors may have a distinct immunological function. It seems likely that the "truth" is a combination of models, paralleling the situation for lymphocytes where both subset commitment and activation play pivotal roles in the development of effector function.

It should be noted that the formal hematopoietic concept of a DC "lineage" is somewhat constrained by the atypical progenitor–terminally differentiated cell relationship (see below) and lack of definitive lineage-specific markers (like TCR rearrangement for T cells). Thus DC are defined by a constellation of somewhat "soft" characteristics, such as morphology, surface marker expression (MHC and costimulatory ligands, etc.), transcription factors (e.g., RelB), cytokine/chemokines expression, and function (crosspresentation of antigen, etc.). However, the functional heterogeneity alone underscores the difficulty of definitive identification of a dendritic cell, and similarly for distinct DC subsets.

3. DC PROGENITORS

DC are unique in the hematopoietic system in that they arise from "convergent" hematopoiesis rather than the traditional "divergent" hematopoiesis (a multipotent stem cell giving rise to increasingly lineage-restricted progenitors). Rather than all arising from the same committed precursor, DC can be generated from a variety of progenitors at different stages of differentiation, ranging from multipotential CD34+ hematopoietic precursor cells (HPC)[38,39], CD34+CD86+ committed bipotential precursors[40], both the common lymphoid progenitors (CLP) and common myeloid progenitors (CMP)[41,42], DC-specific precursors (CFU-DC)[43], terminally differentiated monocytes[44,45], and immature neutrophils[46]. In addition to the untransformed progenitors, hematopoietic malignancies can also be induced to undergo DC differentiation[47-50]. For example, myeloid leukemic blasts across a range of differentiation stages can be driven to differentiate to DC by exogenous cytokines[50,51]. Based on the DC morphology and the differential expression of certain myeloid and lymphoid phenotypic markers, it appears that in vivo both mice and humans contain DC that originated from either lymphoid or myeloid precursors[52], consistent with in vitro findings[43,53-57]. In addition to different precursor cells, culturing the same population of progenitors (such as the CD34+ HPC) in different cytokine cocktails leads to the development of DC subsets, including myeloid interstitial DC and Langerhans cells[58,59] as well as lymphoid-derived DC[53].

4.   WINGED MESSENGERS — SIGNALING PATHWAYS THAT DRIVE DC DIFFERENTIATION

4.1. Extracellular Stimuli

Ex vivo differentiation of DC from both normal and leukemic progenitors can be achieved by both receptor-mediated exogenous stimuli (such as cytokine combinations and CD40 receptor crosslinking) and surface receptor-independent agents that directly activate intracellular signaling pathways (calcium ionophores (CI) that mobilize intracellular calcium $[Ca]_i$, and phorbol esters that activate protein kinase C (PKC))[39,48,60-66]. The cytokines that have been used (in combination) to drive DC differentiation typically include IL-3, SCF, Flt3L, IL-6, GM-CSF, TNF-α, IL-4, and IL-1β. GM-CSF is included in most cytokine combinations that target the generation of DC from myeloid progenitors (but not from CLP[42]) but does not appear essential for generation of DC in vivo, as DC can develop in GM-CSF[-/-] x GM-CSFR[-/-] mice[67]. More recent data have demonstrated that Flt3 ligand (Flt3L) (in the absence of GM-CSF) appears to also play a central role in differentiation of bone marrow (BM) derived DC and has been shown to expand DC precursors both in vitro and in vivo[68-71]. The growth and survival of BM-derived Flt3L-induced DC is augmented by the addition of

IL-6[71]. Stem cell factor (SCF), an important cytokine in early-stage hematopoiesis, also greatly enhances recovery and long-term expansion of DC[70,72].

Along with the soluble mediators of DC differentiation, the CD40 signaling pathway has also emerged as an important factor for DC function. CD40 is expressed on both non-hematopoietic and hematopoietic cells and is upregulated on maturing DC[73]. In addition to its role in potently activating immature DC, CD40 is also functional on CD34⁺ progenitor cells, and its crosslinking by CD154 can, independent of GM-CSF, cause their proliferation and differentiation into potent allo-stimulatory DC[64]. Recombinant CD154-Ig, as a single agent, can also be used to generate DC from monocytes[74].

Similar to the non-transformed progenitors, primary acute myelogenous leukemia (AML) and chronic myelogenous leukemia (CML) isolates can also be driven to undergo DC differentiation using cytokine combinations containing GM-CSF, TNF-α, and IL-4[48,49,66,75-77]. However, since great variability exists within the specific maturation stages at which leukemic cells are arrested (and leukemic clones frequently undergo events that make them refractory to some differentiating agents), additional components are often added to the standard mixture of cytokines used for DC differentiation from normal precursors[78]. In this respect, it has been shown that, depending on a leukemic clone, cytokine combinations containing GM-CSF, IL-4, TNF-α, can be supplemented with SCF, TGFβ, IFN-α, or CD154 to facilitate DC differentiation[49,65,78,79].

4.2. Intracellular Signal Transduction

Intracellular calcium flux, G-protein signaling, protein kinase C (PKC) activation, translocation of NFκB to the nucleus, stimulation of MAPK/ERK, as well as activation of certain transcription factors such as STAT3, PU.1, Notch, and Ikaros[35,36,80,81-83] have all been implicated in various models of DC differentiation. For this review we will focus on PKC activation and downstream pathways.

4.2.1. Protein Kinase C

The ability of different progenitors at different stages of maturation to commit to the DC lineage suggests that DC differentiation signals converge at a common level of intracellular signal transduction pathways. Although a myriad of intracellular signals emanate from cytokine receptors, we have found that protein kinase C (PKC) appears to play a central role in DC differentiation. PKC was first identified by Nishizuka and colleagues as a cyclic nucleotide-independent serine/threonine protein kinase that was calcium (Ca^{2+}) activated and phospholipid dependent[84]. They soon demonstrated that its activity was greatly enhanced by small amounts of diacylglycerol (DAG), a byproduct of the hydrolysis of phosphatidylinositol (4,5)-bisphosphate, establishing for the first time the role of phospholipids as second messengers [85]. The synthesis of relatively hydrophilic phorbol esters like PMA and phorbol dibutyrate (PDBu), which mimick the structure of DAG, allowed to demonstrate the specific binding of DAG to PKC to induce kinase activity[86]. Furthermore, binding to phorbol esters was shown to result in the rapid redistribution of PKC from the cytosol to the cell membrane, and this translocation has served since as the hallmark of PKC enzymatic activation[87]. It is now clear that PKC are a heterogeneous group of proteins. To date, twelve different PKC isoforms have been identified and clustered into three subfamilies on the basis of their activation requirements[88]. Classical PKC isoform (α, βI, βII, and γ) activity is regulated by Ca^{2+}, DAG, and phospholipids; novel PKC isoforms (δ, ε, η, and θ) are regulated by DAG and phospholipids; while the atypical PKC isoforms (ζ and ι or λ, PKCι is the human ortholog of mouse PKCλ) are insensitive to DAG and Ca^{++} [89]. Other interactions with adaptor proteins are required for translocation to the cell membrane to occur. As a result, treatment with DAG (or phorbol esters) often results in selective recruitment of specific isoforms to the membrane, instead of simply all expressed classic and novel proteins. The idea of adaptor proteins mediating PKC localization and activity was pioneered by Mochly-Rosen and colleagues, and led to the identification of RACK proteins as regulators of PKC signaling[90]. There are other layers of regulation of PKC activity — such as the requirement for serine/threonine/tyrosine phosphorylation to yield a mature, catalytically competent enzyme, and tissue-specific variation in isozyme content[89].

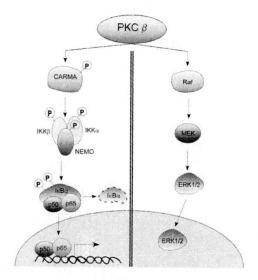

Figure 4. The MAPK/ERK and the NFκB pathways are activated by PKC. PKC-dependent activation of NFκB. The activation of PKC recruits a complex of proteins containing CARMA. This complex of proteins phosphorylates IKK, which then phosphorylates and inactivates IκBα, allowing the NFκB to translocate to the nucleus and regulate the NFκB-dependent gene expression. **PKC-dependent activation of MAPK/ERK.** PKC can activate the ERK pathways through PKC-mediated activation of Raf. Raf, in turn, activates an upstream ERK kinase, MEK, which subsequently phosphorylates ERK, allowing it to translocate to the nucleus to regulate gene expression.

PKC signaling has been reported downstream of a number of key receptors that drive DC differentiation, including CD40[91] and the GM-CSF receptor[92]. PKCβII has also been implicated in the function of Langerhans cells during contact hypersensitivity[93]. We have found that inhibitors of PKC activation block cytokine-mediated DC differentiation from both primary human monocytes and CD34+ HPC[39,94]. Conversely, we have shown that direct activation of PKC by PMA alone induces primary human CD34+ HPC to differentiate directly into DC[39,95]. This PMA-induced DC differentiation occurs without proliferation, is dominant over the generation of other hematopoietic lineages by the cytokine cocktail SCF + IL-3 + GM-CSF + IL-6, and is DC specific as other lineages are not generated. Direct PKC activation drives DC differentiation in myeloid leukemia cell lines[63,96] and primary leukemic blasts[97]. Interestingly, although PMA is a broad PKC activator, we have found a unique role of PKCβII in DC differentiation. Its activation is suggested by the rapid translocation of this isoform, but not other classical isoforms to the plasma membrane post PMA treatment in DC progenitors (see [94] and Lindner et al., manuscript in preparation). Furthermore, the importance of PKCβII is corroborated by the PKCβ siRNA studies that show that downregulation of PKCβ in DC progenitors inhibits DC differentiation. The importance of PKCβII in DC differentiation is corroborated by the findings that overexpression of PKCβII in a differentiation-resistant KG-1a cell line restores its ability to undergo DC differentiation[94]. These studies also suggest that there is complex regulation of PKCβII gene expression that may in turn regulate whether a myeloid precursor can undergo DC differentiation. It also appears that in addition to the requirement for PKC activation, PKC signal strength also controls the downstream pathways such as RelB, and that various levels of PKC activity in cells confer different biological responses[94,98].

Downstream, PKC activation has been previously shown to trigger both the NFκB and MAPK/ERK pathways in other cell types (Figure 4)[99–102]. Both of these pathways have also been implicated in DC differentiation[103,104]. In contrast, activation of the p38 pathway was shown to be detrimental for DC differentiation from monocytes[105,106]. We have found in a cell line model of DC differentiation[96] that PMA induced a rapid (45 min post-treatment) activation of the MAPK/ERK pathway visualized by the phosphorylation of ERK, followed by activation of the NFκB pathway by 24 hours (Lindner et al., in preparation). In contrast to ERK and NFκB, and consistent with previous studies by Xie et al.,[105,106] we also observed a reduction in p38 activity early during PMA-induced DC differentiation, also demonstrating that PMA does not cause global activation of signaling cascades.

4.2.2. MAP Kinase

During DC activation (which, it should again be noted, is distinct from DC differentiation) it has been shown that manipulating the MAPK family of protein kinases (including the MAP kinases extracellular-signal-regulated kinase 1 (ERK1) and ERK2, the c-Jun N-terminal kinase (JNK), and the p38 MAPK) confers different functional properties on the resulting DC[107,108]. It appears that the NFκB, c-Jun, and p38 pathways positively regulate DC maturation by inducing upregulation of CD83, CD86, and CD40, while the ERK pathway may negatively or positively regulate maturation depending on the particular system employed[108–113]. For example, it has been demonstrated that, although ERK is activated during LPS-induced DC maturation[113], its activation inhibits TNF-α-induced maturation of murine DC and TNF-α and LPS-induced maturation of human monocyte-derived DC[110].

We and others[103] have also found that ERK activation plays an important role in DC differentiation. In our studies, pharmacological inhibition of ERK activation during DC differentiation results in reversal of the cell growth arrest/cell death characteristically seen during differentiation, as well as a decreased ability of the resulting DC to activate allogeneic T cells. However, ERK inhibition did not affect upregulation of costimulatory ligand expression (CD40), nor did it have any effect on RelB expression. Since upregulation of RelB is mediated primarily by NFκB signaling (see below), this finding suggests that NFκB signaling is not downstream of ERK activation and that the NFκB and the MAPK/ERK pathways are independently activated by PKC.

4.2.3. NFκB

It is becoming increasingly apparent that distinct subunits of the NFκB transcription factor family are involved in DC differentiation[104,114–117]. Activation of NFκB occurs following phosphorylation by the IKK complex of kinases of the NFκB inhibitory IκB proteins, which results in degradation of IκB and dissociation of the NFκB complex, followed by translocation of NFκB to the nucleus. NFκB exists as a dimer composed of five proteins (p65/RelA, c-Rel, RelB, p50/NFκB1, and p52/NFκB2) and mediates activation of inflammatory and immune-response genes[118–120]. The role of NFκB signaling in DC differentiation and function has been analyzed in some detail. Overexpression of the inhibitory IκB protein in mature DC downregulates the expression of HLA class II, the costimulatory molecules CD80, CD86, and CD40 and the proinflammatory cytokine TNF-α, indicating that antigen presentation is dependant on NFκB function[121]. The effect of individual NFκB proteins in DC development and function has been assessed in chimeric and knockout mice. These studies have shown the generation of functional DC in mice lacking individual p50, p52, RelA, and c-Rel proteins, although the combined deficiency of p50 and RelA results in a severe defect in DC development, and DC from p50$^{-/-}$cRel$^{-/-}$ animals fail to produce IL-12. These data indicate that, although NFκB activity is required for proper differentiation and function of DC, there is some redundancy in the role of individual subunits[122]. On the other hand, the NFκB subunit RelB has a more direct effect in DC development, and is the transcription factor most intensely analyzed in DC differentiation and function[104,123–128].

4.2.3a. RelB

RelB protein can be detected in human and mouse dendritic cells, with high expression in interdigitating DC of thymic medulla and the deep cortex of lymph nodes[129]. Its nuclear expression (as a p50/RelB heterodimer) is one of the hallmarks of DC differentiation and correlates with degree of maturation, including activation of the antigen-presenting capacity[125,126,130]. As a result, RelB-deficient mice show an impaired antigen-presenting cell function and cellular immunity, with a profound decrease of thymic and splenic DC[104]. Experiments in bone marrow chimeric mice demonstrated that the deficiency in splenic DC is an intrinsic defect in the progenitor cells. In addition, antigen-primed DC with inhibited RelB function were shown to lack typical costimulatory molecules and generated antigen-specific regulatory T cells in vivo[26]. Finally, recent studies have shown that RelB promotes differentiation into DC, and its inhibition impairs monocyte-derived DC development with no effect on other myeloid differentiation pathways [131].

RelB has some features that distinguish it from other NFκB members. RelB does not homodimerize and forms heterodimers almost exclusively with p100, p52, and p50 proteins[132,133]. Furthermore, RelB complexes are not bound by IκB, but can be retained in the cytoplasm by p100[134]. Consequently, the inhibitory p100 precursor form acts as a negative regulator of DC function when overexpressed[135]. Processing of p100 to p52 to allow nuclear translocation of the RelB/p52 dimer occurs through a nonclassical signaling pathway that requires IKKα activation by the NFκB-inducing kinase (NIK)[136,137]. In this pathway the precursor p100 in the complex is processed by the proteasome to the p52 form. The precursor p105 also contains IκB-like ankyrin repeats that could sequester NFκB in the cytosol. However, generation of p50 occurs constitutively by a co-translational mechanism from the p105 form[138,139]. As p50 and p52 are expressed in many cells of the immune system, RelB/p50 and RelB/p52 dimers can be found in the nucleus of unstimulated cells[134]. Therefore, while NFκB activity is typically regulated by post-transcriptional events, RelB activity levels in cells expressing p50 and/or p52 proteins parallels increases in RelB transcription[140-143]. The role of RelB in DC differentiation and the fact that its upregulation is a very early event in this process[63,128] suggest that the mechanisms controlling RelB transcription may play an integral role in the differentiation from DC progenitors and in determination of the functional characteristics of the generated DC. How RelB gene expression is regulated, however, has been largely undefined.

Utilizing a cell line model of DC differentiation[63] where the parental cells (KG1) can undergo DC differentiation while a daughter cell line (KG1a) cannot, we have found evidence that RelB gene expression is regulated by PKC signal strength via regulation of both transcription initiation and transcriptional elongation (Figure 5 and [98]). KG1a cells (which have no detectable PKCβII expression) actively initiate transcription from the RelB promoter, but only a fraction of the nascent transcripts can elongate beyond an attenuator element located in intron 4 of the gene. Transcription initiation in KG1 cells, which have low

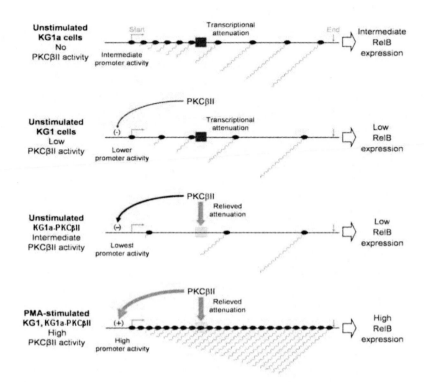

Figure 5. PKCβII controls RelB expression by regulation of transcriptional initiation and elongation. KG1a cells actively initiate transcription from the RelB promoter, but transcription is attenuated. Basal PKCβII activity in untreated KG1 cells downregulates the promoter activity without affecting attenuation, resulting in lower levels of full-length RelB transcripts. Increased expression (and enzyme activity) of PKCβII in untreated KG1a-βII transfectant cells relieves the attenuation, but the stronger downregulation in transcriptional initiation results in even lower levels of RelB mRNA. PMA-mediated activation of PKCβII results in an increase of promoter activity that combines with the relief in attenuation to upregulate RelB mRNA levels.

(basal) levels of PKCβII activity, is lower due to a decrease in p50/RelA protein expression. In addition, the attenuator element is present in untreated KG1 cells and results in further downregulation of RelB expression. In contrast, the higher PKC activity in the PKCβII-transfected KG1a relieves the transcriptional attenuation, allowing transcripts to freely elongate beyond exon 4. However, the higher PKC activity also results in a decreased transcription initiation, ultimately responsible for the observed low levels of RelB mRNA expression. When PKC is fully activated (as in PMA-treated cells) transcriptional elongation occurs uninterrupted beyond exon 4 but now combines with an increase in transcription initiation (due to enhanced levels and nuclear translocation of p50/RelA protein) to upregulate RelB mRNA and protein expression. The proposed model implicates distinct signaling pathways that are selectively triggered at different levels of PKC signaling. Such a model is supported by studies that found different thresholds of PKC activity required for lineage commitment from a hematopoietic precursor cell line model[144].

4.2.3b. Inhibition of NFκB Signaling during DC Differentiation

Previous studies have demonstrated that inhibition of NFκB signaling and RelB upregulation during DC differentiation results in a weakened immune response and generation of regulatory T cells in mice[26], but not in humans[27]. Inhibition of NFκB during murine bone marrow progenitor→DC differentiation results in DC that generate IL-10 secreting regulatory T cells[26], while the same inhibition during human monocyte→DC differentiation yields DC that sensitize (but do not fully activate) T cells[27].

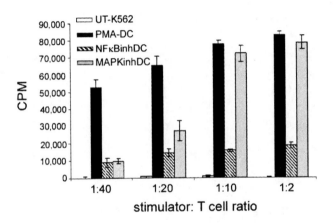

Figure 6. Effect of ERK and NFκB inhibition on allostimulatory ability. Myeloid progenitors were either left undifferentiated, differentiated to DC (with PMA), or differentiated in the presence of an ERK or NFκB inhibitor. These stimulators were then irradiated and cultured at the indicated ratios with purified resting allogeneic T cells. T cell proliferation was assayed by thymidine incorporation.

Using a cell line model of human HSC→DC differentiation, we have found that inhibition of NFκB signaling during differentiation results in enhanced cell proliferation arrest and cell death (suggesting that NFκB plays a role in DC survival), blocks upregulation of CD40, blocks upregulation of RelB expression but enhances activation of ERK signaling (i.e., NFκB negatively regulates the MAPK/ERK pathway), and inhibits the development of allostimulatory capability. However, unlike ERK inhibition, cells differentiated in the presence of NFκB inhibitors were not able to drive greater T cell proliferation in a dose-dependent manner (Figure 6). Further analysis of the T cell responses elicited by the "NFκB-inhibited DC" (NFκBinhDC) demonstrated that even though little proliferation was seen (by thymidine incorporation or CSFE dilution) and the cells became arrested in G1, they upregulated the activation marker CD69 (but not CD25). These T cells could not be restimulated with fresh "normal" DC, nor was IL-2 and γIFN expression induced with either primary or secondary stimulation. Consistent with a lack of CD25 expression, we could find no evidence for suppressive function of sorted T cells in add-back experiments that would suggest that regulatory T cells were being generated by NFκBinhDC. However, T cells activated by the NFκBinhDC could be restimulated by fresh DC in the presence of exogenous IL-2 or agnostic anti-CD28 antibodies, which is characteristic of anergic T cells[145-148]. Furthermore, consistent with the previous observation that CTLA4-mediated anergy is associated with a G1 cell cycle arrest[147], we have found that blocking CTLA4 enhances T cell proliferation induced by NFκBinhDC.

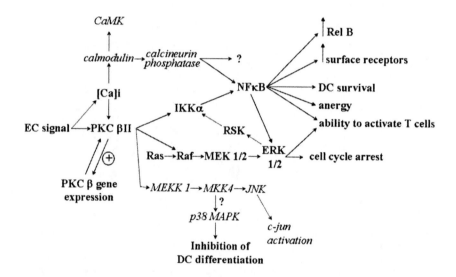

Figure 7. Intracellular signal transduction in DC differentiation.

Together, these findings suggest that different intracellular signaling pathways influence different aspects of DC differentiation and acquisition of specific functions (Figure 7). There is now considerable evidence that these pathways are modulated by a number of exogenous stimuli, including factors overexpressed by tumors (such as VEGF and β2-microglobulin) which inhibit DC differentiation in vitro and in vivo[28,103,149-151] and result in clinically significant immunosuppression in patient. These are, however, likely to be normal immuno-inhibitory mechanisms involved in wound healing (it would not be evolutionarily advantageous to induce robust T cell activation in the setting of sterile trauma) that are magnified in cancer, which is classically described as the "non-healing wound." True pathogenic signal transduction requires pathogens.

5. ACHILLES' HEEL — SUBVERSION OF DC DIFFERENTIATION BY PATHOGEN-MEDIATED DISRUPTION OF SIGNAL TRANSDUCTION PATHWAYS

It is an old adage that if there is a weakness in the immune system, microbial pathogens have figured out a way to exploit it. Given the critical role of DC in initiating adaptive responses, it comes as no surprise that pathogens can subvert host immunity by inactivating DC function. Most of the mechanisms described so far revolve around inhibiting DC activation/maturation. For example, DC exposed to the anthrax lethal toxin (a critical virulence factor of *Bacillus anthracis*) fail to stimulate antigen-specific T cells, and this effect is mediated by inactivation of the MAPK pathway in DC[152]. DC are also targeted by acute infections with viruses such as Ebola and Lassa and impair their function[153].

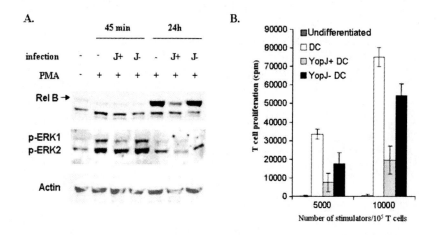

Figure 8. Infection of DC precursors with YopJ-expressing *Y. pestis* strains suppresses their differentiation into DC. (A) YopJ translocation inhibits PMA-induced upregulation of RelB and phospho-ERK1/2. K562 were left uninfected or infected with *Y. pestis poy-Yop* deletion strain lacking all six effector Yops (J-), or *poly-Yop* deletion strain containing the plasmid encoding YopJ-GSK fusion protein (J+). Cells were then cultured for 45 minutes or 24 hours in the absence or presence of PMA. Total proteins were isolated and immunobloted for RelB, phospo-ERK1/2, phospho-GSK or β-actin. **(B) YopJ translocation impairs DC allo-stimulatory capacity.** K562 were left uninfected or infected as in A, and then cultured in the presence or absence of PMA for 5 days. The cells were then irradiated and cocultured for 3 days in triplicate wells allogeneic T cells at the stimulator: T cell ratio indicated.

Success for many pathogens is not only to overcome the initial innate and adaptive immune responses, but to ultimately establish chronic infection in the host, where, ideally, the immune system is tolerant to it but active against other threats against the pathogen's new home. If inhibition of NFκB signaling during myeloid progenitor→DC differentiation yields tolerogenic DC, it seems likely that a pathogen would exploit this. And that pathogen may be *Yersinia*.

5.1. Yersinia and the Disruption of Intracellular Signaling Pathways

Yersinia spp are extracellular Gram-negative bacteria that include *Y. pestis* (the etiologic agent of plague) and the other human pathogenic yersiniae (*Y. enterocolitica* and *Y. pseudotuberculosis*). *Yersinia* can impair host immune resistance by injecting several effector proteins, collectively called *Yersinia* outer proteins (Yops), into the cytoplasm of the host immune cells[154]. The injection of effector Yops (YopE, YopH, YopJ, YopM, YopT, and YpkA) via a Type III secretion system (T3SS)[155,156] results in disruption of the cell's cytoskeleton and inhibition of phagocytosis and inflammation[157-163]. Of particular interest here are the effector Yops known to disrupt host cell signaling pathways, including pathways implicated in DC differentiation and function[164]. Specifically, YopJ, a ubiquitin-like cysteine protease, has been shown to inhibit macrophage function by preventing the activation of MAP kinase kinases (MKKs) and IκB kinase (IKK)-β[160,163,165]. YopJ-dependent disruption of IKKβ monoubiquination prevents degradation of IκBα, an inhibitor of NFκB, and therefore inhibits activation of the NFκB signaling cascade. Furthermore, several reports have suggested that enteropathogenic *Yersinia* spp. inactivate T cell response by inhibiting the function of mature DC[166,167] and inducing DC apoptosis[166,168]. While several groups have demonstrated that DC are injected by *Yersinia* spp. in vitro, a recent study has elegantly verified that *Y. pestis* also preferentially targets DC and macrophages in vivo[169].

5.2. Yersinia and DC Differentiation

Given that *Yersinia* has an effector protein in YopJ that inhibits MAPK and NFκB signaling, and given that it injects this protein into myeloid cells (including monocytes and macrophages) during a natural infection[169,170], it seem possible that YopJ-injected myeloid progenitors that subsequently undergo DC differentiation (possibly in response to the inflammatory milieu) will become tolerogenic DC-presenting *Yersinia* antigens. To begin to assess this, we infected DC progenitors in vitro (both leukemic cell lines and primary human monocytes) with YopJ+ and YopJ– strains of *Y. pestis*, and examined the effect on DC immunogenicity (Lindner et al., in preparation). As seen in Figure 8A, infection with the YopJ+ but not the J– strain results in a fivefold reduction in the induction of RelB expression (as a measure of NFκB signaling), as well as inhibiting ERK 1 and 2 phosphorylation. This was also associated with downregulation of

CD40 expression, consistent with inhibition of NFκB signaling. Functionally, the YopJ⁺-infected, but not YopJ⁻-infected PMA-differentiated DC demonstrated a reduced capacity to stimulate an allogeneic T cell response (Figure 8B).

We have also found that YopJ+ but not YopJ– *Y. pestis* infection of monocytes during GM-CSF + IL-4-induced differentiation similarly inhibited the development of phenotypic, molecular, and functional characteristics of DC. This is consistent with a recent study demonstrating a critical role of YopP (a *Y. enterocolitica*-encoded protein equivalent of YopJ in *Y. pestis*) in suppressing DC function and subsequent T cell response[168]. An important caveat to our findings is that we do not yet know whether YopP is generating tolerogenic DC, but at a minimum it appears that YopJ can significantly disrupt the generation of immunostimulatory DC from myeloid progenitors.

6. CONCLUDING REMARKS

The signal transduction pathways that underlie the ability of diverse progenitors to all differentiate into dendritic cells is only beginning to be defined, but they are likely to play an important role in ultimately shaping the function of the DC. It is becoming clear that specific pathways regulate specific aspects of DC function. And, importantly, these pathways can be manipulated by both endogenous factors and pathogens, and thus may play a central role in both normal and pathogenic DC biology.

7. REFERENCES

1. J. Sprent, H. Kishimoto. The thymus and negative selection. *Immunol Rev* **185**:126–135 (2002).
2. D. Mathis and C. Benoist. Back to central tolerance. *Immunity* **20**(5):509–516 (2004).
3. J. Banchereau, F. Briere, C. Caux, J. Davoust, S. Lebecque, Y.J. Liu, B. Pulendran and K. Palucka. Immunobiology of dendritic cells. *Annu Rev Immunol* **18**(4):767–811 (2000).
4. J. Banchereau and R.M. Steinman. Dendritic cells and the control of immunity. *Nature* **392**(6673):245–252 (1998).
5. M. Rescigno, C. Winzler, D. Delia, C. Mutini, M. Lutz and P. Ricciardi-Castagnoli. Dendritic cell maturation is required for initiation of the immune response. *J Leukoc Biol* **61**(4):415–421 (1997).
6. P. Matzinger. Tolerance, danger, and the extended family. *Annu Rev Immunol* **12**:991–1045 (1994).
7. C. Winzler, P. Rovere, M. Rescigno, F. Granucci, G. Penna, L. Adorini, V.S. Zimmermann, J. Davoust and P. Ricciardi-Castagnoli. Maturation stages of mouse dendritic cells in growth factor-dependent long-term cultures. *J Exp Med* **185**(2):317–328 (1997).

8. E.C. de Jong, H.H. Smits and M.L. Kapsenberg. Dendritic cell-mediated T cell polarization. *Springer Semin Immunopathol* **26**(3):289–307 (2005).

9. K.M. Murphy, W. Ouyang, J.D. Farrar, J. Yang, S. Ranganath, H. Asnagli, M. Afkarian and T.L. Murphy. Signaling and transcription in T helper development. *Annu Rev Immunol* **18**:451–494 (2000).

10. M. Moser and K.M. Murphy. Dendritic cell regulation of TH1–TH2 development. *Nat Immunol* **1**(3):199–205 (2000).

11. A. D'Andrea, X. Ma, M. Aste-Amezaga, C. Paganin and G. Trinchieri. Stimulatory and inhibitory effects of interleukin (IL)-4 and IL-13 on the production of cytokines by human peripheral blood mononuclear cells: priming for IL-12 and tumor necrosis factor alpha production. *J Exp Med* **181**(2):537–546 (1995).

12. C. Heufler, F. Koch, U. Stanzl, G. Topar, M. Wysocka, G. Trinchieri, A. Enk, R.M. Steinman, N. Romani and G. Schuler. Interleukin-12 is produced by dendritic cells and mediates T helper 1 development as well as interferon-gamma production by T helper 1 cells. *Eur J Immunol* **26**(3):659–668 (1996).

13. F. Koch, U. Stanzl, P. Jennewein, K. Janke, C. Heufler, E. Kampgen, N. Romani and G. Schuler. High level IL-12 production by murine dendritic cells: upregulation via MHC class II and CD40 molecules and downregulation by IL-4 and IL-10. *J Exp Med* **184**(2):741–746 (1996).

14. M. Croft, D.D. Duncan and S.L. Swain. Response of naive antigen-specific CD4+ T cells in vitro: characteristics and antigen-presenting cell requirements. *J Exp Med* **176**(5):1431–1437 (1992).

15. S.E. Macatonia, C.S. Hsieh, K.M. Murphy and A. O'Garra. Dendritic cells and macrophages are required for Th1 development of CD4+ T cells from alpha beta TCR transgenic mice: IL-12 substitution for macrophages to stimulate IFN-gamma production is IFN-gamma-dependent. *Int Immunol* **5**(9):1119–1128 (1993).

16. A.H. Enk. Dendritic cells in tolerance induction. *Immunol Lett* **99**(1):8–11 (2005).

17. M.V. Dhodapkar, R.M. Steinman, J. Krasovsky, C. Munz and N. Bhardwaj. Antigen-specific inhibition of effector T cell function in humans after injection of immature dendritic cells. *J Exp Med* **193**(2):233–238 (2001).

18. M.B. Lutz, R.M. Suri, M. Niimi, A.L. Ogilvie, N.A. Kukutsch, S. Rossner, G. Schuler and J.M. Austyn. Immature dendritic cells generated with low doses of GM-CSF in the absence of IL-4 are maturation resistant and prolong allograft survival in vivo. *Eur J Immunol* **30**(7):1813–1822 (2000).

19. H. Jonuleit, E. Schmitt, G. Schuler, J. Knop and A.H. Enk. Induction of interleukin 10-producing, nonproliferating CD4(+) T cells with regulatory properties by repetitive stimulation with allogeneic immature human dendritic cells. *J Exp Med* **192**(9):1213–1222 (2000).

20. K. Steinbrink, M. Wolfl, H. Jonuleit, J. Knop and A.H. Enk. Induction of tolerance by IL-10-treated dendritic cells. *J Immunol* **159**(10):4772–4780 (1997).

21. L. Piemonti, P. Monti, P. Allavena, M. Sironi, L. Soldini, B.E. Leone, C. Socci and V. Di Carlo. Glucocorticoids affect human dendritic cell differentiation and maturation. *J Immunol* **162**(11):6473–6481 (1999).

22. L. Piemonti, P. Monti, M. Sironi, P. Fraticelli, B.E. Leone, E. Dal Cin, P. Allavena and V. Di Carlo. Vitamin D3 affects differentiation, maturation, and function of human monocyte-derived dendritic cells. *J Immunol* **164**(9):4443–4451 (2000).

23. P. Kalinski, C.M. Hilkens, A. Snijders, F.G. Snijdewint and M.L. Kapsenberg. IL-12-deficient dendritic cells, generated in the presence of prostaglandin E2, promote

type 2 cytokine production in maturing human naive T helper cells. *J Immunol* **159**(1):28–35 (1997).

24. A.H. Enk, H. Jonuleit, J. Saloga and J. Knop. Dendritic cells as mediators of tumor-induced tolerance in metastatic melanoma. *Int J Cancer* **73**(3):309–316 (1997).

25. C. Rastellini, L. Lu, C. Ricordi, T.E. Starzl, A.S. Rao and A.W. Thomson. Granulo-cyte/macrophage colony-stimulating factor-stimulated hepatic dendritic cell progeni-tors prolong pancreatic islet allograft survival. *Transplantation* **60**(11):1366–1370 (1995).

26. E. Martin, B. O'Sullivan, P. Low and R. Thomas. Antigen-specific suppression of a primed immune response by dendritic cells mediated by regulatory T cells secreting interleukin-10. *Immunity* **18**(1):155–167 (2003).

27. A.G. Thompson, B.J. O'Sullivan, H. Beamish and R. Thomas. T cells signaled by NF-kappa B- dendritic cells are sensitized not anergic to subsequent activation. *J Immunol* **173**(3):1671–1680 (2004).

28. D. Gabrilovich, T. Ishida, T. Oyama, S. Ran, V. Kravtsov, S. Nadaf and D.P. Car-bone. Vascular endothelial growth factor inhibits the development of dendritic cells and dramatically affects the differentiation of multiple hematopoietic lineages in vivo. *Blood* **92**(11):4150–4166 (1998).

29. K. Shortman and Y.J. Liu. Mouse and human dendritic cell subtypes. *Nat Rev Im-munol* **2**(3):151–161 (2002).

30. Q. Huang, D. Liu, P. Majewski, L.C. Schulte, J.M. Korn, R.A. Young, E.S. Lander and N. Hacohen. The plasticity of dendritic cell responses to pathogens and their components. *Science* **294**(5543):870–875 (2001).

31. Y.J. Liu. Dendritic cell subsets and lineages, and their functions in innate and adap-tive immunity. *Cell* **106**(3):259–262 (2001).

32. R. Maldonado-Lopez, T. De Smedt, P. Michel, J. Godfroid, B. Pajak, C. Heirman, K. Thielemans, O. Leo, J. Urbain and M. Moser. CD8alpha+ and CD8alpha– sub-classes of dendritic cells direct the development of distinct T helper cells in vivo. *J Exp Med* **189**(3):587–592 (1999).

33. B. Pulendran, J.L. Smith, G. Caspary, K. Brasel, D. Pettit, E. Maraskovsky and C.R. Maliszewski. Distinct dendritic cell subsets differentially regulate the class of im-mune response in vivo. *Proc Natl Acad Sci USA* **96**(3):1036–1041 (1999).

34. I. Szatmari, P. Gogolak, J.S. Im, B. Dezso, E. Rajnavolgyi and L. Nagy. Activation of PPARgamma specifies a dendritic cell subtype capable of enhanced induction of iNKT cell expansion. *Immunity* **21**(1):95–106 (2004).

35. P. Cheng, Y. Nefedova, L. Miele, B.A. Osborne and D. Gabrilovich. Notch signal-ing is necessary but not sufficient for differentiation of dendritic cells. *Blood* **102**(12):3980–3988 (2003).

36. K.L. Anderson, H. Perkin, C.D. Surh, S. Venturini, R.A. Maki and B.E. Torbett. Transcription factor PU.1 is necessary for development of thymic and myeloid pro-genitor-derived dendritic cells. *J Immunol* **164**(4):1855–1861 (2000).

37. B. Almand, J.I. Clark, E. Nikitina, J. van Beynen, N.R. English, S.C. Knight, D.P. Carbone and D.I. Gabrilovich. Increased production of immature myeloid cells in cancer patients: a mechanism of immunosuppression in cancer. *J Immunol* **166**(1):678–689 (2001).

38. C. Caux, C. Dezutter-Dambuyant, D. Schmitt and J. Banchereau. GM-CSF and TNF-alpha cooperate in the generation of dendritic Langerhans cells. *Nature* **360**(6401):258–261 (1992).

39. T.A. Davis, A.A. Saini, P.J. Blair, B.L. Levine, N. Craighead, D.M. Harlan, C.H. June and K.P. Lee, Phorbol esters induce differentiation of human CD34+ hemopoietic progenitors to dendritic cells: evidence for protein kinase C-mediated signaling. *J Immunol* **160**(8):3689–3697 (1998).

40. R.E. Ryncarz and C. Anasetti. Expression of CD86 on human marrow CD34(+) cells identifies immunocompetent committed precursors of macrophages and dendritic cells. *Blood* **91**(10):3892–3900 (1998).

41. M.G. Manz, D. Traver, K. Akashi, M. Merad, T. Miyamoto, E.G. Engleman and I.L. Weissman. Dendritic cell development from common myeloid progenitors. *Ann NY Acad Sci* **938**(167–173; discussion 173–164) (2001).

42. M.G. Manz, D. Traver, T. Miyamoto, I.L. Weissman and K. Akashi. Dendritic cell potentials of early lymphoid and myeloid progenitors. *Blood* **97**(11):3333–3341 (2001).

43. J.W. Young, P. Szabolcs and M.A. Moore. Identification of dendritic cell colony-forming units among normal human CD34+ bone marrow progenitors that are expanded by c-kit-ligand and yield pure dendritic cell colonies in the presence of granulocyte/macrophage colony-stimulating factor and tumor necrosis factor alpha. *J Exp Med* **182**(4):1111–1119 (1995).

44. S.M. Kiertscher and M.D. Roth. Human CD14+ leukocytes acquire the phenotype and function of antigen-presenting dendritic cells when cultured in GM-CSF and IL-4. *J Leukoc Biol* **59**(2):208–218 (1996).

45. L.J. Zhou and T.F. Tedder. CD14+ blood monocytes can differentiate into functionally mature CD83+ dendritic cells. *Proc Natl Acad Sci USA* **93**(6):2588–2592 (1996).

46. L. Oehler, O. Majdic, W.F. Pickl, J. Stockl, E. Riedl, J. Drach, K. Rappersberger, K. Geissler and W. Knapp. Neutrophil granulocyte-committed cells can be driven to acquire dendritic cell characteristics. *J Exp Med* **187**(7):1019–1028 (1998).

47. B.C. Hulette, G. Rowden, C.A. Ryan, C.M. Lawson, S.M. Dawes, G.M. Ridder and G.F. Gerberick. Cytokine induction of a human acute myelogenous leukemia cell line (KG-1) to a CD1a+ dendritic cell phenotype. *Arch Dermatol Res* **293**(3):147–158 (2001).

48. A. Cignetti, E. Bryant, B. Allione, A. Vitale, R. Foa and M.A. Cheever. CD34(+) acute myeloid and lymphoid leukemic blasts can be induced to differentiate into dendritic cells. *Blood* **94**(6):2048–2055 (1999).

49. B.A. Choudhury, J.C. Liang, E.K. Thomas, L. Flores-Romo, Q.S. Xie, K. Agusala, S. Sutaria, I. Sinha, R.E. Champlin and D.F. Claxton. Dendritic cells derived in vitro from acute myelogenous leukemia cells stimulate autologous, antileukemic T-cell responses. *Blood* **93**(3):780–786 (1999).

50. A. Charbonnier, B. Gaugler, D. Sainty, M. Lafage-Pochitaloff and D. Olive. Human acute myeloblastic leukemia cells differentiate in vitro into mature dendritic cells and induce the differentiation of cytotoxic T cells against autologous leukemias. *Eur J Immunol* **29**(8):2567–2578 (1999).

51. B.D. Harrison, J.A. Adams, M. Briggs, M.L. Brereton and J.A. Yin. Stimulation of autologous proliferative and cytotoxic T-cell responses by "leukemic dendritic cells" derived from blast cells in acute myeloid leukemia. *Blood* **97**(9):2764–2771 (2001).

52. N. Bendriss-Vermare, C. Barthelemy, I. Durand, C. Bruand, C. Dezutter-Dambuyant, N. Moulian, S. Berrih-Aknin, C. Caux, G. Trinchieri and F. Briere.

Human thymus contains IFN-alpha-producing CD11c(–), myeloid CD11c(+), and mature interdigitating dendritic cells. *J Clin Invest* **107**(7):835–844 (2001).

53. A. Galy, M. Travis, D. Cen and B. Chen. Human T, B, natural killer, and dendritic cells arise from a common bone marrow progenitor cell subset. *Immunity* **3**(4):459–473 (1995).

54. F. Sallusto and A. Lanzavecchia. Efficient presentation of soluble antigen by cultured human dendritic cells is maintained by granulocyte/macrophage colony-stimulating factor plus interleukin 4 and downregulated by tumor necrosis factor alpha. *J Exp Med* **179**(4):1109–1118 (1994).

55. M.C. Rissoan, V. Soumelis, N. Kadowaki, G. Grouard, F. Briere, R. de Waal Malefyt and Y.J. Liu. Reciprocal control of T helper cell and dendritic cell differentiation. *Science* **283**(5405):1183–1186 (1999).

56. G.J. Randolph, S. Beaulieu, S. Lebecque, R.M. Steinman and W.A. Muller. Differentiation of monocytes into dendritic cells in a model of transendothelial trafficking. *Science* **282**(5388):480–483 (1998).

57. G. Grouard, M.C. Rissoan, L. Filgueira, I. Durand, J. Banchereau and Y.J. Liu. The enigmatic plasmacytoid T cells develop into dendritic cells with interleukin (IL)-3 and CD40-ligand. *J Exp Med* **185**(6):1101–1111 (1997).

58. C. Caux, C. Massacrier, B. Vanbervliet, B. Dubois, B. de Saint-Vis, C. Dezutter-Dambuyant, C. Jacquet, D. Schmitt and J. Banchereau. CD34+ hematopoietic progenitors from human cord blood differentiate along two independent dendritic cell pathways in response to GM-CSF+TNF alpha. *Adv Exp Med Biol* **417**(5):21–25 (1997).

59. C. Caux, B. Vanbervliet, C. Massacrier, B. Dubois, C. Dezutter-Dambuyant, D. Schmitt and J. Banchereau. Characterization of human CD34+ derived dendritic/Langerhans cells (D-Lc). *Adv Exp Med Biol* **378**:1–5 (1995).

60. M. Waclavicek, A. Berer, L. Oehler, J. Stockl, E. Schloegl, O. Majdic and W. Knapp. Calcium ionophore: a single reagent for the differentiation of primary human acute myelogenous leukaemia cells towards dendritic cells. *Br J Haematol* **114**(2):466–473 (2001).

61. B.J. Czerniecki, C. Carter, L. Rivoltini, G.K. Koski, H.I. Kim, D.E. Weng, J.G. Roros, Y.M. Hijazi, S. Xu, S.A. Rosenberg and P.A. Cohen. Calcium ionophore-treated peripheral blood monocytes and dendritic cells rapidly display characteristics of activated dendritic cells. *J Immunol* **159**(8):3823–3837 (1997).

62. G.K. Koski, G.N. Schwartz, D.E. Weng, B.J. Czerniecki, C. Carter, R.E. Gress and P.A. Cohen. Calcium mobilization in human myeloid cells results in acquisition of individual dendritic cell-like characteristics through discrete signaling pathways. *J Immunol* **163**(1):82–92 (1999).

63. D.C. St Louis, J.B. Woodcock, G. Fransozo, P.J. Blair, L.M. Carlson, M. Murillo, M.R. Wells, A.J. Williams, D.S. Smoot, S. Kaushal, J.L. Grimes, D.M. Harlan, J.P. Chute, C.H. June, U. Siebenlist and K.P. Lee. Evidence for distinct intracellular signaling pathways in CD34+ progenitor to dendritic cell differentiation from a human cell line model. *J Immunol* **162**(6):3237–3248 (1999).

64. L. Flores-Romo, P. Bjorck, V. Duvert, C. van Kooten, S. Saeland and J. Banchereau. CD40 ligation on human cord blood CD34+ hematopoietic progenitors induces their proliferation and differentiation into functional dendritic cells. *J Exp Med* **185**(2):341–349 (1997).

65. L. Oehler, A. Berer, M. Kollars, F. Keil, M. Konig, M. Waclavicek, O. Haas, W. Knapp, K. Lechner and K. Geissler. Culture requirements for induction of dendritic cell differentiation in acute myeloid leukemia. *Ann Hematol* **79**(7):355–362 (2000).

66. M. Heinzinger, C.F. Waller, A. von den Berg, A. Rosenstiel and W. Lange. Generation of dendritic cells from patients with chronic myelogenous leukemia. *Ann Hematol* **78**(4):181–186 (1999).

67. D. Vremec and K. Shortman. Dendritic cell subtypes in mouse lymphoid organs: cross-correlation of surface markers, changes with incubation, and differences among thymus, spleen, and lymph nodes. *J Immunol* **159**(2):565–573 (1997).

68. E. Maraskovsky, K. Brasel, M. Teepe, E.R. Roux, S.D. Lyman, K. Shortman and H.J. McKenna. Dramatic increase in the numbers of functionally mature dendritic cells in Flt3 ligand-treated mice: multiple dendritic cell subpopulations identified. *J Exp Med* **184**(5):1953–1962 (1996).

69. B. Pulendran, J. Lingappa, M.K. Kennedy, J. Smith, M. Teepe, A. Rudensky, C.R. Maliszewski and E. Maraskovsky. Developmental pathways of dendritic cells in vivo: distinct function, phenotype, and localization of dendritic cell subsets in FLT3 ligand-treated mice. *J Immunol* **159**(5):2222–2231 (1997).

70. A. Curti, M. Fogli, M. Ratta, S. Tura and R.M. Lemoli. Stem cell factor and FLT3-ligand are strictly required to sustain the long-term expansion of primitive CD34+DR– dendritic cell precursors. *J Immunol* **166**(2):848–854 (2001).

71. J. Taieb, K. Maruyama, C. Borg, M. Terme and L. Zitvogel, Imatinib mesylate impairs Flt3L-mediated dendritic cell expansion and antitumor effects in vivo. *Blood* **103**(5):1966–1967 [author reply, 1967] (2004).

72. K. Saraya and C.D. Reid. Stem cell factor and the regulation of dendritic cell production from CD34+ progenitors in bone marrow and cord blood. *Br J Haematol* **93**(2):258–264 (1996).

73. S. Akira, K. Takeda and T. Kaisho. Toll-like receptors: critical proteins linking innate and acquired immunity. *Nat Immunol* **2**(8):675–680 (2001).

74. G.M. Zou and Y.K. Tam, Cytokines in the generation and maturation of dendritic cells: recent advances. *Eur Cytokine Network* **13**(2):186–199 (2002).

75. B. Eibl, S. Ebner, C. Duba, G. Bock, N. Romani, M. Erdel, A. Gachter, D. Niederwieser and G. Schuler. Dendritic cells generated from blood precursors of chronic myelogenous leukemia patients carry the Philadelphia translocation and can induce a CML-specific primary cytotoxic T-cell response. *Genes Chromosomes Cancer* **20**(3):215–223 (1997).

76. A. Choudhury, A. Toubert, S. Sutaria, D. Charron, R.E. Champlin and D.F. Claxton. Human leukemia-derived dendritic cells: ex-vivo development of specific antileukemic cytotoxicity. *Crit Rev Immunol* **18**(1–2):121–131 (1998).

77. F. Santiago-Schwarz, D.L. Coppock, A.A. Hindenburg and J. Kern. Identification of a malignant counterpart of the monocyte-dendritic cell progenitor in an acute myeloid leukemia. *Blood* **84**(9):3054–3062 (1994).

78. C. Wang, H.M. Al-Omar, L. Radvanyi, A. Banerjee, D. Bouman, J. Squire and H.A. Messner. Clonal heterogeneity of dendritic cells derived from patients with chronic myeloid leukemia and enhancement of their T-cells stimulatory activity by IFN-alpha. *Exp Hematol* **27**(7):1176–1184 (1999).

79. A. Cignetti, A. Vallario, I. Roato, P. Circosta, B. Allione, L. Casorzo, P. Ghia and F. Caligaris-Cappio. Leukemia-derived immature dendritic cells differentiate into func-

tionally competent mature dendritic cells that efficiently stimulate T cell responses. *J Immunol* **173**(4):2855–2865 (2004).

80. Y. Laouar, T. Welte, X.Y. Fu and R.A. Flavell. STAT3 is required for Flt3L-dependent dendritic cell differentiation. *Immunity* **19**(6):903–912 (2003).

81. L. Wu, A. Nichogiannopoulou, K. Shortman and K. Georgopoulos. Cell-autonomous defects in dendritic cell populations of Ikaros mutant mice point to a developmental relationship with the lymphoid lineage. *Immunity* **7**(4):483–492 (1997).

82. Y. Laouar, I.N. Crispe and R.A. Flavell. Overexpression of IL-7R alpha provides a competitive advantage during early T-cell development. *Blood* **103**(6):1985–1994 (2004).

83. R. Schotte, M.C. Rissoan, N. Bendriss-Vermare, J.M. Bridon, T. Duhen, K. Weijer, F. Briere and H. Spits. The transcription factor Spi-B is expressed in plasmacytoid DC precursors and inhibits T-, B-, and NK-cell development. *Blood* **101**(3):1015–1023 (2003).

84. Y. Takai, A. Kishimoto, Y. Iwasa, Y. Kawahara, T. Mori and Y. Nishizuka. Calcium-dependent activation of a multifunctional protein kinase by membrane phospholipids. *J Biol Chem* **254**(10):3692–3695 (1979).

85. Y. Takai, A. Kishimoto, U. Kikkawa, T. Mori and Y. Nishizuka. Unsaturated diacylglycerol as a possible messenger for the activation of calcium-activated, phospholipid-dependent protein kinase system. *Biochem Biophys Res Commun* **91**(4):1218–1224 (1979).

86. M. Castagna, Y. Takai, K. Kaibuchi, K. Sano, U. Kikkawa and Y. Nishizuka. Direct activation of calcium-activated, phospholipid-dependent protein kinase by tumor-promoting phorbol esters. *J Biol Chem* **257**(13):7847–7851 (1982).

87. A.S. Kraft and W.B. Anderson. Phorbol esters increase the amount of Ca2+, phospholipid-dependent protein kinase associated with plasma membrane. *Nature* **301**(5901):621–623 (1983).

88. M. Spitaler and D.A. Cantrell. Protein kinase C and beyond. *Nat Immunol* **5**(8):785–790 (2004).

89. W.S. Liu and C.A. Heckman. The sevenfold way of PKC regulation. *Cell Signal* **10**(8):529–542 (1998).

90. D. Schechtman and D. Mochly-Rosen. Adaptor proteins in protein kinase C-mediated signal transduction. *Oncogene* **20**(44):6339–6347 (2001).

91. J.C. Yan, Z.G. Wu, X.T. Kong, R.Q. Zong and L.Z. Zhang. Effect of CD40-CD40 ligand interaction on diacylglycerol-protein kinase C signal transduction pathway and intracellular calcium in cultured human monocytes. *Acta Pharmacol Sin* **24**(7):687–691 (2003).

92. N. Geijsen, M. Spaargaren, J.A. Raaijmakers, J.W. Lammers, L. Koenderman and P.J. Coffer, Association of RACK1 and PKCbeta with the common beta-chain of the IL-5/IL-3/GM-CSF receptor. *Oncogene* **18**(36):5126–5130 (1999).

93. A.L. Goodell, H.S. Oh, S.A. Meyer and R.C. Smart. Epidermal protein kinase C-beta 2 is highly sensitive to downregulation and is exclusively expressed in Langerhans cells: downregulation is associated with attenuated contact hypersensitivity. *J Invest Dermatol* **107**(3):354–359 (1996).

94. P.J. Cejas, L.M. Carlson, J. Zhang, S. Padmanabhan, D. Kolonias, I. Lindner, S. Haley, L.H. Boise and K.P. Lee. Protein kinase C betaII plays an essential role in

dendritic cell differentiation and autoregulates its own expression. *J Biol Chem* **280**(31):28412–28423 (2005).

95. G. Ramadan, R.E. Schmidt and J. Schubert. In vitro generation of human CD86+ dendritic cells from CD34+ haematopoietic progenitors by PMA and in serum-free medium. *Clin Exp Immunol* **125**(2):237–244 (2001).

96. I. Lindner, M.A. Kharfan-Dabaja, E. Ayala, D. Kolonias, L.M. Carlson, Y. Beazer-Barclay, U. Scherf, J.H. Hnatyszyn and K.P. Lee, Induced dendritic cell differentiation of chronic myeloid leukemia blasts is associated with down-regulation of BCR-ABL. *J Immunol* **171**(4):1780–1791 (2003).

97. M. Kharfan-Dabaja, E. Ayala, I. Lindner, P.J. Cejas, N.J. Bahlis, D. Kolonias, L.M. Carlson and K.P. Lee. Differentiation of acute and chronic myeloid leukemic blasts into the dendritic cell lineage: analysis of various differentiation-inducing signals. *Cancer Immunol Immunother* **54**(1):25–36 (2005).

98. P.J. Cejas, L.M. Carlson, D. Kolonias, J. Zhang, I. Lindner, D.D. Billadeau, L.H. Boise and K.P. Lee. Regulation of RelB expression during the initiation of dendritic cell differentiation. *Mol Cell Biol* **25**(17):7900–7916 (2005).

99. A. Cariappa, L. Chen, K. Haider, M. Tang, E. Nebelitskiy, S.T. Moran and S. Pillai. A catalytically inactive form of protein kinase C-associated kinase/receptor interacting protein 4, a protein kinase C beta-associated kinase that mediates NF-kappa B activation, interferes with early B cell development. *J Immunol* **171**(4):1875–1880 (2003).

100. A.K. Olsson, K. Vadhammar and E. Nanberg. Activation and protein kinase C-dependent nuclear accumulation of ERK in differentiating human neuroblastoma cells. *Exp Cell Res* **256**(2):454–467 (2000).

101. T.T. Su, B. Guo, Y. Kawakami, K. Sommer, K. Chae, L.A. Humphries, R.M. Kato, S. Kang, L. Patrone, R. Wall, M. Teitell, M. Leitges, T. Kawakami and D.J. Rawlings. PKC-beta controls I kappa B kinase lipid raft recruitment and activation in response to BCR signaling. *Nat Immunol* **3**(8):780–786 (2002).

102. A. Kumar, T.C. Chambers, B.A. Cloud-Heflin and K.D. Mehta. Phorbol ester-induced low density lipoprotein receptor gene expression in HepG2 cells involves protein kinase C-mediated p42/44 MAP kinase activation. *J Lipid Res* **38**(11):2240–2248 (1997).

103. J. Xie, Y. Wang, M.E. Freeman 3rd, B. Barlogie and Q. Yi. Beta 2-microglobulin as a negative regulator of the immune system: high concentrations of the protein inhibit in vitro generation of functional dendritic cells. *Blood* **101**(10):4005–4012 (2003).

104. L. Wu, A. D'Amico, K.D. Winkel, M. Suter, D. Lo and K. Shortman. RelB is essential for the development of myeloid-related CD8alpha– dendritic cells but not of lymphoid-related CD8alpha+ dendritic cells. *Immunity* **9**(6):839–847 (1998).

105. J. Xie, J. Qian, S. Wang, M.E. Freeman 3rd, J. Epstein and Q. Yi. Novel and detrimental effects of lipopolysaccharide on in vitro generation of immature dendritic cells: involvement of mitogen-activated protein kinase p38. *J Immunol* **171**(9):4792–4800 (2003).

106. J. Xie, J. Qian, J. Yang, S. Wang, M.E. Freeman 3rd and Q. Yi. Critical roles of Raf/MEK/ERK and PI3K/AKT signaling and inactivation of p38 MAP kinase in the differentiation and survival of monocyte-derived immature dendritic cells. *Exp Hematol* **33**(5):564–572 (2005).

107. W. Lim, W. Ma, K. Gee, S. Aucoin, D. Nandan, F. Diaz-Mitoma, M. Kozlowski and A. Kumar. Distinct role of p38 and c-Jun N-terminal kinases in IL-10-dependent and

IL-10-independent regulation of the costimulatory molecule B7.2 in lipopolysaccharide-stimulated human monocytic cells. *J Immunol* **168**(4):1759–1769 (2002).

108. K. Sato, H. Nagayama, K. Tadokoro, T. Juji and T.A. Takahashi. Extracellular signal-regulated kinase, stress-activated protein kinase/c-Jun N-terminal kinase, and p38mapk are involved in IL-10-mediated selective repression of TNF-alpha-induced activation and maturation of human peripheral blood monocyte-derived dendritic cells. *J Immunol* **162**(7):3865–3872 (1999).

109. J.F. Arrighi, M. Rebsamen, F. Rousset, V. Kindler and C. Hauser. A critical role for p38 mitogen-activated protein kinase in the maturation of human blood-derived dendritic cells induced by lipopolysaccharide, TNF-alpha, and contact sensitizers. *J Immunol* **166**(6):3837–3845 (2001).

110. A. Puig-Kroger, M. Relloso, O. Fernandez-Capetillo, A. Zubiaga, A. Silva, C. Bernabeu and A.L. Corbi. Extracellular signal-regulated protein kinase signaling pathway negatively regulates the phenotypic and functional maturation of monocyte-derived human dendritic cells. *Blood* **98**(7):2175–2182 (2001).

111. A. Puig-Kroger, F. Sanz-Rodriguez, N. Longo, P. Sanchez-Mateos, L. Botella, J. Teixido, C. Bernabeu and A.L. Corbi. Maturation-dependent expression and function of the CD49d integrin on monocyte-derived human dendritic cells. *J Immunol* **165**(8):4338–4345 (2000).

112. Y. Yanagawa, N. Iijima, K. Iwabuchi and K. Onoe. Activation of extracellular signal-related kinase by TNF-alpha controls the maturation and function of murine dendritic cells. *J Leukoc Biol* **71**(1):125–132 (2002).

113. M. Rescigno, M. Martino, C.L. Sutherland, M.R. Gold and P. Ricciardi-Castagnoli. Dendritic cell survival and maturation are regulated by different signaling pathways. *J Exp Med* **188**(11):2175–2180 (1998).

114. S. Yoshimura, J. Bondeson, F.M. Brennan, B.M. Foxwell and M. Feldmann. Role of NFkappaB in antigen presentation and development of regulatory T cells elucidated by treatment of dendritic cells with the proteasome inhibitor PSI. *Eur J Immunol* **31**(6):1883–1893 (2001).

115. S. Yoshimura, J. Bondeson, B.M. Foxwell, F.M. Brennan and M. Feldmann. Effective antigen presentation by dendritic cells is NF-kappaB dependent: coordinate regulation of MHC, co-stimulatory molecules and cytokines. *Int Immunol* **13**(5):675–683 (2001).

116. F. Ouaaz, J. Arron, Y. Zheng, Y. Choi and A.A. Beg. Dendritic cell development and survival require distinct NF-kappaB subunits. *Immunity* **16**(2):257–270 (2002).

117. A.R. Pettit, K.P. MacDonald, B. O'Sullivan and R. Thomas. Differentiated dendritic cells expressing nuclear RelB are predominantly located in rheumatoid synovial tissue perivascular mononuclear cell aggregates. *Arthr Rheum* **43**(4):791–800 (2000).

118. P.A. Baeuerle and T. Henkel. Function and activation of NF-kappa B in the immune system. *Annu Rev Immunol* **12**(141–1790 (1994).

119. S. Ghosh, M.J. May and E.B. Kopp. NF-kappa B and Rel proteins: evolutionarily conserved mediators of immune responses. *Annu Rev Immunol* **16**:225–260 (1998).

120. F.G. Wulczyn, D. Krappmann and C. Scheidereit. The NF-kappa B/Rel and I kappa B gene families: mediators of immune response and inflammation. *J Mol Med* **74**(12):749–769 (1996).

121. S. Yoshimura, J. Bondeson, B.M. Foxwell, F.M. Brennan and M. Feldmann. Effective antigen presentation by dendritic cells is NF-kappaB dependent: coordinate

regulation of MHC, co-stimulatory molecules and cytokines. *Int Immunol* 13(5):675–683 (2001).

122. F. Ouaaz, J. Arron, Y. Zheng, Y. Choi and A.A. Beg. Dendritic cell development and survival require distinct NF-kappaB subunits. *Immunity* 16(2):257–270 (2002).

123. D.S. Weih, Z.B. Yilmaz and F. Weih. Essential role of RelB in germinal center and marginal zone formation and proper expression of homing chemokines. *J Immunol* 167(4):1909–1919 (2001).

124. R. Valero, M.L. Baron, S. Guerin, S. Beliard, H. Lelouard, B. Kahn-Perles, B. Vialettes, C. Nguyen, J. Imbert and P. Naquet. A defective NF-kappa B/RelB pathway in autoimmune-prone New Zealand black mice is associated with inefficient expansion of thymocyte and dendritic cells. *J Immunol* 169(1):185–192 (2002).

125. B.J. O'Sullivan, K.P. MacDonald, A.R. Pettit and R. Thomas. RelB nuclear translocation regulates B cell MHC molecule, CD40 expression, and antigen-presenting cell function. *Proc Natl Acad Sci USA* 97(21):11421–11426 (2000).

126. G.J. Clark, S. Gunningham, A. Troy, S. Vuckovic and D.N. Hart. Expression of the RelB transcription factor correlates with the activation of human dendritic cells. *Immunology* 98(2):189–196 (1999).

127. F. Weih, G. Warr, H. Yang and R. Bravo. Multifocal defects in immune responses in RelB-deficient mice. *J Immunol* 158(11):5211–5218 (1997).

128. M. Neumann, H. Fries, C. Scheicher, P. Keikavoussi, A. Kolb-Maurer, E. Brocker, E. Serfling and E. Kampgen. Differential expression of Rel/NF-kappaB and octamer factors is a hallmark of the generation and maturation of dendritic cells. *Blood* 95(1):277–285 (2000).

129. D. Carrasco, R.P. Ryseck and R. Bravo. Expression of relB transcripts during lymphoid organ development: specific expression in dendritic antigen-presenting cells. *Development* 118(4):1221–1231 (1993).

130. A.R. Pettit, K.P. MacDonald, B. O'Sullivan and R. Thomas. Differentiated dendritic cells expressing nuclear RelB are predominantly located in rheumatoid synovial tissue perivascular mononuclear cell aggregates. *Arthr Rheum* 43(4):791–800 (2000).

131. B. Platzer, A. Jorgl, S. Taschner, B. Hocher and H. Strobl. RelB regulates human dendritic cell subset development by promoting monocyte intermediates. *Blood* 104(12):3655–3663 (2004).

132. R.P. Ryseck, P. Bull, M. Takamiya, V. Bours, U. Siebenlist, P. Dobrzanski and R. Bravo. RelB, a new Rel family transcription activator that can interact with p50-NF-kappa B. *Mol Cell Biol* 12(2):674–684 (1992).

133. P. Dobrzanski, R.P. Ryseck and R. Bravo. Specific inhibition of RelB/p52 transcriptional activity by the C-terminal domain of p100. *Oncogene* 10(5):1003–1007 (1995).

134. N.J. Solan, H. Miyoshi, E.M. Carmona, G.D. Bren and C.V. Paya. RelB cellular regulation and transcriptional activity are regulated by p100. *J Biol Chem* 277(2):1405–1418 (2002).

135. K. Speirs, L. Lieberman, J. Caamano, C.A. Hunter and P. Scott. Cutting edge: NF-kappa B2 is a negative regulator of dendritic cell function. *J Immunol* 172(2):752–756 (2004).

136. G. Bonizzi and M. Karin. The two NF-kappaB activation pathways and their role in innate and adaptive immunity. *Trends Immunol* 25(6):280–288 (2004).

137. G. Xiao, E.W. Harhaj and S.C. Sun. NF-kappaB-inducing kinase regulates the processing of NF-kappaB2 p100. *Mol Cell* 7(2):401–409 (2001).

138. M. Karin and Y. Ben-Neriah. Phosphorylation meets ubiquitination: the control of NF-[kappa]B activity. *Annu Rev Immunol* **18**:621–663 (2000).

139. L. Lin, G.N. DeMartino and W.C. Greene. Cotranslational biogenesis of NF-kappaB p50 by the 26S proteasome. *Cell* **92**(6):819–828 (1998).

140. T. Lernbecher, B. Kistler and T. Wirth. Two distinct mechanisms contribute to the constitutive activation of RelB in lymphoid cells. *Embo J* **13**(17):4060–4069 (1994).

141. D.A. Francis, R. Sen, N. Rice and T.L. Rothstein. Receptor-specific induction of NF-kappaB components in primary B cells. *Int Immunol* **10**(3):285–293 (1998).

142. B. Kistler, A. Rolink, R. Marienfeld, M. Neumann and T. Wirth. Induction of nuclear factor-kappa B during primary B cell differentiation. *J Immunol* **160**(5):2308–2317 (1998).

143. M. Suhasini, C.D. Reddy, E.P. Reddy, J.A. DiDonato and R.B. Pilz. cAMP-induced NF-kappaB (p50/relB) binding to a c-myb intronic enhancer correlates with c-myb up-regulation and inhibition of erythroleukemia cell differentiation. *Oncogene* **15**(15):1859–1870 (1997).

144. F. Rossi, M. McNagny, G. Smith, J. Frampton and T. Graf. Lineage commitment of transformed haematopoietic progenitors is determined by the level of PKC activity. *Embo J* **15**(8):1894–1901 (1996).

145. S. Glennie, I. Soeiro, P.J. Dyson, E.W. Lam and F. Dazzi. Bone marrow mesenchymal stem cells induce division arrest anergy of activated T cells. *Blood* **105**(7):2821–2827 (2005).

146. B. Verdoodt, T. Blazek, P. Rauch, G. Schuler, A. Steinkasserer, M.B. Lutz and J.O. Funk. The cyclin-dependent kinase inhibitors p27Kip1 and p21Cip1 are not essential in T cell anergy. *Eur J Immunol* **33**(11):3154–3163 (2003).

147. A.D. Wells, M.C. Walsh, J.A. Bluestone and L.A. Turka. Signaling through CD28 and CTLA-4 controls two distinct forms of T cell anergy. *J Clin Invest* **108**(6):895–903 (2001).

148. J. Sun, M. Alison Stalls, K.L. Thompson and N. Fisher Van Houten. Cell cycle block in anergic T cells during tolerance induction. *Cell Immunol* **225**(1):33–41 (2003).

149. D.I. Gabrilovich, H.L. Chen, K.R. Girgis, H.T. Cunningham, G.M. Meny, S. Nadaf, D. Kavanaugh and D.P. Carbone. Production of vascular endothelial growth factor by human tumors inhibits the functional maturation of dendritic cells. *Nat Med* **2**(10):1096–1103 (1996).

150. T. Oyama, S. Ran, T. Ishida, S. Nadaf, L. Kerr, D.P. Carbone and D.I. Gabrilovich. Vascular endothelial growth factor affects dendritic cell maturation through the inhibition of nuclear factor-kappa B activation in hemopoietic progenitor cells. *J Immunol* **160**(3):1224–1232 (1998).

151. N. Boissel, P. Rousselot, E. Raffoux, J.M. Cayuela, O. Maarek, D. Charron, L. Degos, H. Dombret, A. Toubert and D. Rea. Defective blood dendritic cells in chronic myeloid leukemia correlate with high plasmatic VEGF and are not normalized by imatinib mesylate. *Leukemia* **18**(10):1656–1661 (2004).

152. A. Agrawal, J. Lingappa, S.H. Leppla, S. Agrawal, A. Jabbar, C. Quinn and B. Pulendran. Impairment of dendritic cells and adaptive immunity by anthrax lethal toxin. *Nature* **424**(6946):329–334 (2003).

153. S. Mahanty, K. Hutchinson, S. Agarwal, M. McRae, P.E. Rollin and B. Pulendran. Cutting edge: impairment of dendritic cells and adaptive immunity by Ebola and Lassa viruses. *J Immunol* **170**(6):2797–2801 (2003).

154. G. Denecker, W. Declercq, C.A. Geuijen, A. Boland, R. Benabdillah, M. van Gurp, M.P. Sory, P. Vandenabeele and G.R. Cornelis. *Yersinia enterocolitica* YopP-induced apoptosis of macrophages involves the apoptotic signaling cascade up-stream of bid. *J Biol Chem* **276**(23):19706–19714 (2001).

155. C.J. Hueck. Type III protein secretion systems in bacterial pathogens of animals and plants. *Microbiol Mol Biol Rev* **62**(2):379–433 (1998).

156. L.E. Palmer, A.R. Pancetti, S. Greenberg and J.B. Bliska. YopJ of Yersinia spp. is sufficient to cause downregulation of multiple mitogen-activated protein kinases in eukaryotic cells. *Infect Immun* **67**(2):708–716 (1999).

157. I. Sorg, U.M. Goehring, K. Aktories and G. Schmidt. Recombinant Yersinia YopT leads to uncoupling of RhoA-effector interaction. *Infect Immun* **69**(12):7535–7543 (2001).

158. M. Iriarte and G.R. Cornelis. YopT, a new Yersinia Yop effector protein, affects the cytoskeleton of host cells. *Mol Microbiol* **29**(3):915–929 (1998).

159. R. Rosqvist, A. Forsberg and H. Wolf-Watz. Intracellular targeting of the Yersinia YopE cytotoxin in mammalian cells induces actin microfilament disruption. *Infect Immun* **59**(12):4562–4569 (1991).

160. K. Schesser, A.K. Spiik, J.M. Dukuzumuremyi, M.F. Neurath, S. Pettersson and H. Wolf-Watz. The yopJ locus is required for Yersinia-mediated inhibition of NF-kappaB activation and cytokine expression: YopJ contains a eukaryotic SH2-like domain that is essential for its repressive activity. *Mol Microbiol* **28**(6):1067–1079 (1998).

161. G.R. Cornelis, A. Boland, A.P. Boyd, C. Geuijen, M. Iriarte, C. Neyt, M.P. Sory and I. Stainier. The virulence plasmid of Yersinia, an antihost genome. *Microbiol Mol Biol Rev* **62**(4):1315–1352 (1998).

162. M. Aepfelbacher, R. Zumbihl, K. Ruckdeschel, C.A. Jacobi, C. Barz and J. Heese-mann. The tranquilizing injection of Yersinia proteins: a pathogen's strategy to resist host defense. *Biol Chem* **380**(7–8):795–802 (1999).

163. K. Ruckdeschel, S. Harb, A. Roggenkamp, M. Hornef, R. Zumbihl, S. Kohler, J. Heesemann and B. Rouot. Yersinia enterocolitica impairs activation of transcription factor NF-kappaB: involvement in the induction of programmed cell death and in the suppression of the macrophage tumor necrosis factor alpha production. *J Exp Med* **187**(7):1069–1079 (1998).

164. K. Orth, Z. Xu, M.B. Mudgett, Z.Q. Bao, L.E. Palmer, J.B. Bliska, W.F. Mangel, B. Staskawicz and J.E. Dixon. Disruption of signaling by Yersinia effector YopJ, a ubiquitin-like protein protease. *Science* **290**(5496):1594–1597 (2000).

165. K. Orth, L.E. Palmer, Z.Q. Bao, S. Stewart, A.E. Rudolph, J.B. Bliska and J.E. Dixon. Inhibition of the mitogen-activated protein kinase kinase superfamily by a Yersinia effector. *Science* **285**(5435):1920–1923 (1999).

166. S.E. Erfurth, S. Grobner, U. Kramer, D.S. Gunst, I. Soldanova, M. Schaller, I.B. Autenrieth and S. Borgmann. Yersinia enterocolitica induces apoptosis and inhibits surface molecule expression and cytokine production in murine dendritic cells. *Infect Immun* **72**(12):7045–7054 (2004).

167. M. Schoppet, A. Bubert and H.I. Huppertz. Dendritic cell function is perturbed by *Yersinia enterocolitica* infection in vitro. *Clin Exp Immunol* **122**(3):316–323 (2000).

168. K. Trulzsch, G. Geginat, T. Sporleder, K. Ruckdeschel, R. Hoffmann, J. Heesemann and H. Russmann. Yersinia outer protein P inhibits CD8 T cell priming in the mouse infection model. *J Immunol* **174**(7):4244–4251 (2005).

169. M.M. Marketon, R.W. Depaolo, K.L. Debord, B. Jabri and O. Schneewind. Plague bacteria target immune cells during infection. *Science* **309**(5741):1739-1741 (2005).
170. B.A. Chromy:J. Perkins, J.L. Heidbrink, A.D. Gonzales, G.A. Murphy, J.P. Fitch and S.L. McCutchen-Maloney. Proteomic characterization of host response to *Yersinia pestis* and near neighbors. *Biochem Biophys Res Commun* **320**(2):474–479 (2004).

SHAPING NAIVE AND MEMORY CD8+ T CELL RESPONSES IN PATHOGEN INFECTIONS THROUGH ANTIGEN PRESENTATION

Gabrielle T. Belz, Nicholas S. Wilson, Fiona Kupresanin,
Adele M. Mount, and Christopher M. Smith

1. INTRODUCTION

The phenotypic and functional studies carried out during recent years have highlighted the enormous heterogeneity among dendritic cells. These specialized cells possess a variety of features that make them highly efficient agents for the detection of pathogens and induction of immune responses. Unraveling how the phenotypic, molecular, and functional signatures of dendritic cells regulate the decision-making process during an immune response has been the focus of intense research in recent years. The advances in our understanding have implications for the development of vaccine strategies that are targeted to individual subpopulations of dendritic cells.

2. DENDRITIC CELLS OF SPLEEN AND LYMPH NODES

Dendritic cells (DC) are professional antigen-presenting cells (APC) that have an extraordinary capacity to stimulate naive T cells and initiate primary

Gabrielle T. Belz, Adele Mount, and Fiona Kupresanin, Division of Immunology, The Walter and Eliza Hall Institute of Medical Research, Melbourne, 3050 Australia. Nicholas S. Wilson, CSL Limited, Melbourne, 3052 Australia. Christopher M. Smith, Department of Pathology, University of Cambridge, Cambridge, CB2 2QQ UK. Address correspondence to: Gabrielle Belz, belz@wehi.edu.au   Tel. +61 (03) 9345 2534, fax +61 3 9347 0852.

immune responses to pathogens. They are continuously generated in the bone marrow and are widely distributed as immature DC to both lymphoid and non-lymphoid tissues[1].

Gaining an understanding of the origins and development of DC has proven difficult. This is attributable to their rarity in lymphoid tissues (<1%). Both common lymphoid and common myeloid progenitors appear to have the capacity to differentiate into the different subsets of conventional DC suggesting that the DC lineage has incredible developmental flexibility[2,3]. While some reports have described the conversion of one DC type to another[4-6], other studies have failed to find strong evidence to support such a developmental relationship. Most of our information on DC behavior and their classification are derived from studies examining DC phenotype and function in their steady-state environment[7]. However, it is likely that these details may need to be modified as we better understand the enormous plasticity of DC in effectively shaping an immune response to pathogens.

The existence of multiple DC subsets with distinct microenvironmental niches points toward unique functional specialization of different DC. In the spleen and lymph nodes (LN) up to seven subsets of DC that express intermediate to high levels of the integrin CD11c have been described[8-13]. One of these subsets corresponds to the IFNα-producing plasmacytoid DC (pDC)[11,12]. The three conventional DC populations found in the spleen can be distinguished using the surface markers CD4, CD8α, CD11b, and CD205 (Table 1). One subset expresses CD8α together with CD205, but lacks expression of CD11b (CD8 DC). Another subset expresses CD4 and CD11b, but not CD205 (CD4 DC), while a third subset expresses only CD11b. This latter subset is referred to as the double-negative (DN) DC. A fourth subset of DC found in lymph node (LN) but not in spleen expresses both CD205 and CD11b. This subset is the equivalent of the "interstitial" DC found in many peripheral tissues. In the skin they are referred to as dermal DC, while an equivalent DC subset dominant in the LN draining the lung (interstitial-like), but also found in the hepatic and renal LN and Peyer's patch, expresses CD205 but not CD8α or CD11b[8,14]. In addition to dermal DC are those DC resident only in the epidermis of the skin, the Langerhans cells.

Langerhans cells that have migrated to the draining LN express CD11b, CD205, and low or negligible amounts of CD8α. The current paradigm suggests that interstitial, interstitial-like DC, and Langerhans cells ("tissue-derived DC") carry antigens from peripheral tissues to the draining LN, where they present them to other lymphoid cells. This contrasts with the three conventional DC subsets found in spleen and LN that do not appear to traffic from peripheral tissues prior to entering the secondary lymphoid tissues. Rather, they appear to be best defined as "resident DC," which have originated from the bone marrow precursors that seed secondary lymphoid tissues via the blood[7].

Table 1. Conventional DC Subsets in Mouse Lymphoid Tissues

| | DC subsets | | | | Tissue distribution** | | | |
| | Surface marker* | | | | | Mes-enteric | Skin-draining | Visceral |
Subset designation	CD8	CD4	CD205	CD11b	Spleen	LN	LN	LN
CD8 DC	+++	–	++	–	++	+	+	+
CD4 DC	–	+++	–	+++	++++	+/–	+/–	+/–
DN DC	–	–	–	+	++	+++	++	++
Dermal/ interstitial DC	–	–	+	+	+/–	++	++	+
Langerhans cells	+	–	+++	+	–	+/–	+	–
Interstitial-like DC***	–		++	–	–	+	+	+++

*   The relative level of expression of each surface marker on DC subsets.
** The relative frequency of DC subsets is expressed by the number of '+' symbols: 50–70% (++++), 30–50% (+++), 20–30% (++), 10–20% (+), and <5% (+/–).

3. ROLE OF DENDRITIC CELLS IN PATHOGEN RESPONSES

3.1. Priming Naive T Cells

Classically, priming of naive CD8$^+$ T cells requires professional APC that can efficiently present endogenous or pathogen-derived antigens on major histo-compatibility (MHC) class I molecules in combination with the necessary costimulatory molecules to facilitate full activation of T cells. However, whether each of these elements is strictly required to be provided by a professional APC to enable T cell priming has remained contentious. Pathogens usually provide an abundant source of antigen together with pathogen-derived components (for example, cell wall lipids) that are themselves highly inflammatory. Together, such signals may to be sufficient for many cells of the body that express MHC class I molecules to activate T cells, thereby sidestepping the absolute require-ment for professional APC.

Early studies by Staerz and collegues[15] showed that, following influenza in-fection, mice failed to develop virus-specific CD8$^+$ T cells when phagocytic cells were depleted in vivo, but that priming was restored when macrophages were administered. This provided the first direct evidence that phagocytic cells play an important role in priming CD8$^+$ T cell responses to viral infection. Sometime later Rock and colleagues[16,17] exploited the sensitivity of most bone marrow-derived cells to irradiation to demonstrate that bone marrow-derived cells are essential for virus-specific CD8⁺ T cell priming. In these studies irradiated C57BL/6 recipient mice were transplanted with bone marrow from Tap1$^{o/o}$ mice. This type of bone marrow lacks the transporter required for presentation of immunogenic peptides to CD8⁺ T cells. Analogous approaches have utilized bone marrow cells that carry mutant MHC class I molecules that are unable to present immunogenic peptide from the antigen under examination. In these systems, only parenchymal cells (non-bone marrow-derived cells) would be able to present antigens to CD8⁺ T cells. Such an approach has been used to examine the response to vaccinia virus, lymphocytic choriomeningitis virus, and influenza virus[17,18]. Collectively these studies have elegantly demonstrated that virus-specific naive CD8⁺ T cell responses require antigen to be presented on MHC class I molecules by bone marrow-derived cells.

While the above experiments have established that bone marrow-derived cells are generally required to elicit pathogen-specific responses, it has proven far more difficult to elucidate (i) whether presentation of virus-derived antigens is strictly limited to bone marrow-derived professional APC, and (ii) if so, what is the exact identity of these APC. The approach to the first problem was pioneered by Debrick et al.[15], and more recently refined by Jung et al.[19], who designed an elegant transgenic mouse model to eliminate CD11c⁺ cells. The latter group developed transgenic mice that express the diphtheria toxin receptor fused to the green fluorescent protein (DTR-GFP) driven by the CD11c promotor. Mouse cells do not naturally express the diphtheria toxin receptor and thus CD11c⁺ cells become susceptible to the cytotoxic effects of diphtheria toxin, allowing inducible ablation of DC in vivo. Strikingly, mice that were depleted of DC failed to develop T cell responses following either malaria, *Listeria monocytogenes,* or viral infections, confirming the crucial importance of DC in initiating naive CD8⁺ T cell responses to pathogens[19,20].

The second problem described above, that of defining the specific identity of DC actually presenting the pathogen-derived antigens, has been facilitated by several laboratories, including our own, developing very careful methods for direct ex vivo DC purification and analysis. These approaches have been used to mainly examine pathogen systems and will be described in greater detail below.

3.2. Identifying the Main Movers and Shakers in Infection

DC show amazing phenotypic diversity, resulting in the many different subsets described above. Such diversity raises the notion that, like T cells and B cells, DC subsets represent specialized populations of immune cells that respond to different types of antigens or pathogens. This tantalizing concept has fueled an extensive search for DC subtypes that might differentially regulate the induction of T cell immunity or, alternately, T cell tolerance, in vivo[8,21-27].

Two experimental approaches have been used to examine which DC subsets are essential for the CD8$^+$ T cell response to a number of pathogens. Importantly, these experiments hinge on developing highly sensitive in-vitro assays that allow the monitoring of antigen presentation by the very low number of DC thought to carry pathogen-derived antigens. Both approaches involve inoculating mice with a pathogen and at various time points after infection the draining LN or spleen are dissociated and the DC subsets isolated. These highly purified populations are then co-cultured with T cell hybridoma lines or naive transgenic T cells. In the first experimental system, purified DC are co-cultured with a T cell hybridoma line specific for an MHC class I-restricted peptide derived from the pathogen. Conventionally, the response elicited has been analyzed by measuring interleukin-2 production. However, this approach has rarely proven sufficiently sensitive in pathogen systems to accurately analyze antigen presentation. This further supports the notion that only very small numbers of DC actually carry the viral antigen of interest. More recently, the difficulty in detecting MHC class I presentation directly ex vivo from pathogen-infected animals has been circumvented by using T cell hybridomas that express the *lacZ* gene. This feature enables individual cells that have been stimulated by antigen-bearing cells to become blue on exposure to β-galactosidase[8,21,28]. The advantage of T cell hybridomas is that they are independent of costimulatory requirements and therefore provide a highly sensitive readout for analyzing antigen presentation ex vivo. Interestingly, however, not every cell type that is able to stimulate a *lacZ*-expressing hybridoma can activate naive T cells. To explore which cells have the capacity to fully signal a naive T cell, 5,6-carboxy-fluorescein succinyl ester (CFSE)-labeled naive T cell receptor transgenic cell proliferation has been used. This technique has been central to understanding which cells are essential for the T cell-APC interaction that leads to immunity. Nevertheless, differences in the sensitivity of the respective TCR transgenic cells together with a lack of TCR transgenic T cells for both CD4$^+$ and CD8$^+$ T cell epitopes, which would permit MHC class II antigen presentation to be monitored simultaneously, has left significant gaps in our understanding of how the pathogen response is molded.

3.3. Dendritic Cell Subsets in Pathogen Infections

DC are crucial in mounting an effective cytotoxic T cell response to both lymphocytic choriomeningitis virus (LCMV) and the bacterium *Listeria monocytogenes*[19,29,30]. This most likely involves a complex interplay of different DC populations, encompassing not only conventional DC but also plasmacytoid DC and novel subsets such as Tip DC[29,31,32]. In the case of LCMV, the influx of IFNα-producing pDC limits viral replication[33]. In *Listeria monocytogenes* infection, Tip DC (CD11c$^+$CD11b$^+$ DC), so named for their production of tumor necrosis factor-α (TNFα) and inducible nitric oxide synthase (iNOS), provide cytokine-directed innate control of infection[31]. Although both pDC and Tip DC can present pathogen-derived antigens, presentation is very inefficient when compared to conventional DC and was not required for generation of antigen-specific adaptive immunity[31].

It is likely that the route of foreign antigen invasion into the body is a major factor determining which DC and other APC are involved in the transport of antigen to the LN and, in some instances, the subsequent transferral of antigen within the lymphoid organs to secondary DC. Access to secondary lymphoid tissues typically occurs either via the blood (for example, malaria, yellow fever, and lyme disease) or by transport from peripheral sites such as the lung, gut, or skin. Classically, tissue-derived DC have been implicated as central for initiating immunity. However, more recent rigorous examination of this concept has illustrated that the migratory tissue-derived DC are not always able to prime naive T cells. For example, our group has recently demonstrated that Langerhans cells in mice were unable to prime naive CD8[+] T cells following cutaneous infection with herpes simplex 1 (HSV-1)[26].

Further analysis of the differential roles of DC subsets in priming naive virus-specific CD8[+] T cells has been undertaken in our laboratory during recent years[8,24-26,34]. In particular, one subset of DC, the CD8α DC, appears to play an integral role in presenting virus-specific antigens to naive T cells during infection. We have shown that virus administered via the blood results in MHC class I antigen presentation solely by this conventional CD8α DC subset. Moreover, this was also the case for other viruses, including HSV-1, influenza virus, vaccinia virus, LCMV, and *L. monocytogenes*[25], suggesting that CD8α DC represent a common pathway for handling and presenting pathogen-derived antigens. These studies, however, did not examine the role of DC subsets in CD4[+] T cell presentation following infection. In contrast to the central role played by CD8α DC, CD11b[-] (interstitial) DC seem to be largely responsible for priming naive CD4[+] T cells. For example, Filippi et al.[35] showed that following *Leishmania major* infection CD11b[-]DC presented the MHC class II-restricted LACK antigen to CD4[+] T cells. Similarly, analysis of MHC class II presentation to CD4[+] T cells during herpes simplex 2 (HSV-2) infection revealed CD11b[-]DC as the main APC in draining LN[36]. Interestingly, Langerhans cells did not prime CD4[+] T cells following either *L. major* or HSV-2 infection. In our own studies of antigen presentation to CD8[+] T cells following HSV-1 infection, neither Langerhans cells nor dermal DC could activate naive T cells despite the abundance of antigen associated with both these DC at the infection site.

Our study examining CD8[+] T cell immunity to viral lung infection revealed evidence for the interplay between the migratory tissue-derived DC (CD11b[-] CD8α[-]) and LN resident blood-derived CD8α DC in priming naive CD8[+] T cells[8]. To track DC migration from the lung, CFSE was administered intranasally following virus infection. This dye labeled peripheral lung DC and allowed us to establish that MHC class I antigen presentation was accomplished by the tissue-derived migratory CD11b[-]CD8[-]DC subset. In addition, the LN resident blood-derived CD8α DC efficiently presented viral antigens; however, given that they were not labeled by CFSE, it is most likely that they acquired antigen from the immigrant tissue-derived DC.

Table 2. Dendritic Cell Populations Associated with Pathogen Infections

Infection	DC subset	Naïve T cell activated	Origin/ function	Reference
HSV-1 (cutaneous, subcutaneous)	CD8α[+]	CD8[+]	blood-derived Ag presentation	24–26
HSV-2 (vaginal)	CD11b[+]	CD4[+]	tissue-derived Ag presentation	36
Influenza A, HSV-1 (intranasal)	CD8α[+]	CD8[+]	blood-derived Ag presentation	8 and unpub. (GTB)
	CD11b[+]	CD8[+]	tissue-derived Ag presentation	
LCMV, vaccinia virus (intravenous, intraperitoneal)	CD8α[+]	CD8[+]	blood-derived Ag presentation	25,37
Reovirus	CD8α[+]	CD4[+]	blood-derived Ag presentation	14
	CD11b[low]	CD4[+]	tissue-derived Ag presentation	
Listeria monocytogenes	CD8α[+]	CD8[+]	blood-derived Ag presentation	37
	TipDC	…	iNOS antigen presentation	
Leishmania major	CD11b[+] (dermal DC) CD8α[+]	CD4[+]	tissue-derived Ag presentation	39

*Presentation to CD4[+] T cells has not been examined. **Presentation to CD8[+] T cells has not been examined. ***T cell subset has not been determined.

In a model of reovirus infection of the gut, Fleeton et al.[14] similarly identified two populations of DC involved in generating CD4[+] T cell responses to the virus. These were the CD8α DC and CD11b[low] DC subsets. This latter subset appears to be analogous to the CD11b[-]DC found in the lung and visceral LN both in phenotype and function, and they were found to be important for transporting apoptotic material from the gut epithelium to the mesenteric LN for transfer to resident CD8α DC. These two studies underline the importance of the migratory tissue-derived DC in transporting viral antigens to the draining LN for transfer to LN-resident DC as a generalized mechanism for amplifying the CD8[+] T cell immune response.

3.4. Amplification of Memory CD8+ T Cells in Secondary Infections

A number of studies have identified the ability of various APC — such as DC, macrophages, or even epithelial cells — to differentially present viral antigens. Such work has led to the proposal of an elegant model in which during a primary immune response DC are the essential drivers of T cell priming. In contrast, during a recall response, non-DC, particularly those that are tissue resident, would be ideally positioned to rapidly amplify memory T cells ("tissue mediated")[40]. This would presumably provide the most efficient mechanism for removing infection at the site of entry. From a teleological perspective, such a schema could explain how tissue-mediated antigen presentation influences and facilitates the memory T cell response in vivo. However, it would be extremely important to first determine whether the memory T cells, like naive T cells, in fact depend only on DC to drive their development and differentiation. The current concept that memory T cells could be activated by parenchymal cells is supported by findings showing that memory T cells have a lower threshold for activation and have less stringent costimulation requirements than naive T cells[41-43]. This would argue that memory T cells are more promiscuous than naive T cells in responding to antigen presented by non-DC. To formally address this issue, our group[27] and Zammit et al.[20] established complementary systems in which it was examined whether a bone marrow-derived cell, or specifically a DC, was required to activate and amplify memory CD8[+] T cells in pathogen infections. Remarkably, both studies showed that memory CD8[+] T cells were largely dependent on DC to maximize the recall response to infection. Furthermore, this amplification of memory T cells was reliant on the migratory DC transporting antigen to the draining LN. However, some amplification of effector cells was detected in the lung bronchoalveolar lavage following influenza infection, supporting that non-DC may represent a *bone fide* cell type capable of stimulating memory T cells allowing a tissue-mediated frontline defense against pathogen invasion[20,27]. Interestingly, though, a similar outcome was not apparent when infection was transmitted via a cutaneous route with HSV-1[27]. These studies provide an important new conceptual viewpoint of memory T cells, showing that the interaction with DC is a major mechanism driving both naive T cell activation and memory T cell reactivation.

4. CONCLUSIONS

The heterogeneity and complexity of the DC network have been probed through many meticulous studies over recent years. From this body of work has emerged a picture showing a complex and dynamic interplay between DC subsets and other immune cells. Although much remains to be unraveled about the precise details about the DC subsets and molecular mechanisms regulating DC interactions with T cells, two important observations have emerged that challenge our previous views of the behavior of these cells. The first salient conclusion of the studies presented here is that the induction of T cell immunity to peripheral pathogens requires the tissue-resident migratory DC to transport antigen captured in the periphery to the draining lymph node, where transfer of antigen to other DC can occur. This transfer appears essential to initiating and amplifying T cell immunity, as unexpectedly not all migratory DC (for example, Langerhans cells) are themselves able to stimulate robust T cell responses. Second, memory T cells share with naive T cells a significant dependence on DC to initiate the recall response to pathogens in vivo. This surprising finding raises important questions about how memory T cells are regulated in vivo and how we can harness their features to best recall them in secondary immune responses. Understanding the complex interactions between DC subsets and other immune cells has important implications for the development of targeted vaccine strategies that target specific DC populations in vivo.

5. ACKNOWLEDGMENTS

Our work is supported by grants from the National Health and Medical Research Council (Aust) (G.T.B) and the Wellcome Trust Foundation (UK) (G.T.B). G.T.B. is a Wellcome Trust Senior Overseas Fellow and a Howard Hughes Medical Institute International Scholar. C.M.S. is a CJ Martin NHMRC International Fellow.

6. REFERENCES

1. J. Banchereau and R.M. Steinman. Dendritic cells and the control of immunity. *Nature* **392**:245–252 (1998).
2. L. Wu, A. D'Amico, H. Hochrein, M. O'Keeffe, K. Shortman and K. Lucas. Development of thymic and splenic dendritic cell populations from different hemopoietic precursors. *Blood* **98**:3376–3382 (2001).
3. M.G. Manz, D. Traver, T. Miyamoto, I.L. Weissman and K. Akashi. Dendritic cell potentials of early lymphoid and myeloid progenitors. *Blood* **97**:3333–3341 (2001).
4. G. Moron, P. Rueda, I. Casal and C. Leclerc. CD8aCD11b⁺ dendritic cells present exogenous virus-like particles to CD8⁺ T cells and subsequently express CD8aand CD205 molecules. *J Exp Med* **195**:1233–1245 (2002).

5.   G. Martinez del Hoyo, P. Martin, C.F. Arias, A.R. Marin and C. Ardavin. CD8a⁺ dendritic cells originate from the CD8a⁻ dendritic cell subset by a maturation process involving CD8a, DEC-205, and CD24 up-regulation. *Blood* **99**:999–1004 (2002).

6.   E.I. Zuniga, D.B. McGavern, J.L. Pruneda-Paz, C. Teng and M.B. Oldstone. Bone marrow plasmacytoid dendritic cells can differentiate into myeloid dendritic cells upon virus infection. *Nat Immunol* **5**:1227–1234 (2004).

7.   N.S. Wilson and J.A. Villadangos. Lymphoid organ dendritic cells: beyond the Langerhans cells paradigm. *Immunol Cell Biol* **82**:91–98 (2004).

8.   G.T. Belz, C.M. Smith, L. Kleinert, P. Reading, A. Brooks, K. Shortman, F.R. Carbone, Heath WR. Distinct migrating and nonmigrating dendritic cell populations are involved in MHC class I-restricted antigen presentation after lung infection with virus. *Proc Natl Acad Sci USA* **101**:8670–8675 (2004).

9.   D. Vremec, J. Pooley, H. Hochrein, L. Wu and K. Shortman. CD4 and CD8 expression by dendritic cell subtypes in mouse thymus and spleen. *J Immunol* **164**:2978–2986 (2000).

10. S. Henri, D. Vremec, A. Kamath, J. Waithman, S. Williams, C. Benoist, K. Burnham, S. Saeland, E. Handman, K. Shortman. The dendritic cell populations of mouse lymph nodes. *J Immunol* **167**:741–748 (2001).

11. M. O'Keeffe, H. Hochrein, D. Vremec, I. Caminschi, J.L. Miller, E.M. Anders, L. Wu, M.H. Lahoud, S. Henri, B. Scott, P. Hertzog, L. Tatarczuch, K. Shortman. Mouse plasmacytoid cells: long-lived cells, heterogeneous in surface phenotype and function, that differentiate into CD8⁺ dendritic cells only after microbial stimulus. *J Exp Med* **196**:1307–1319 (2002).

12. M. O'Keeffe, H. Hochrein, D. Vremec, B. Scott, P. Hertzog, L. Tatarczuch and K. Shortman. Dendritic cell precursor populations of mouse blood: identification of the murine homologues of human blood plasmacytoid pre-DC2 and CD11c⁺ DC1 precursors. *Blood* **101**:1453–1459 (2003).

13. F.P. Siegal, N. Kadowaki, M. Shodell, P.A. Fitzgerald-Bocarsly, K. Shah, S. Ho, S. Antonenko, Y.J. Liu. The nature of the principal type 1 interferon-producing cells in human blood. *Science* **284**:1835–1837 (1999).

14. M.N. Fleeton, N. Contractor, F. Leon, J.D. Wetzel, T.S. Dermody and B.L. Kelsall. Peyer's patch dendritic cells process viral antigen from apoptotic epithelial cells in the intestine of reovirus-infected mice. *J Exp Med* **200**:235–245 (2004).

15. J.E. Debrick, P.A. Campbell and U.D. Staerz. Macrophages as accessory cells for class I MHC-restricted immune responses. *J Immunol* **147**:2846–2851 (1991).

16. L.J. Sigal, S. Crotty, R. Andino and K.L. Rock. Cytotoxic T-cell immunity to virus-infected non-haematopoietic cells requires presentation of exogenous antigen. *Nature* **398**:77–80 (1999).

17. L.J. Sigal and K.L. Rock. Bone marrow-derived antigen-presenting cells are required for the generation of cytotoxic T lymphocyte responses to viruses and use transporter associated with antigen presentation (TAP)-dependent and -independent pathways of antigen presentation. *J Exp Med* **192**:1143–1150 (2000).

18. L.L. Lenz, E.A. Butz and M.J. Bevan. Requirements for bone marrow-derived antigen-presenting cells in priming cytotoxic T cell responses to intracellular pathogens. *J Exp Med* **192**:1135–1142 (2000).

19. S. Jung, D. Unutmaz, P. Wong, G. Sano, K. De los Santos, T. Sparwasser, S. Wu, S. Vuthoori, K. Ko. F. Zavala, E.G. Pamer, D.R. Littman and R.A. Lang. In vivo de-

pletion of CD11c⁺ dendritic cells abrogates priming of CD8⁺ T cells by exogenous cell-associated antigens. *Immunity* **17**:211–220 (2002).

20. D.J. Zammit, L.S. Cauley, Q.M. Pham and L. Lefrancois. Dendritic cells maximize the memory CD8 T cell response to infection. *Immunity* **22**:561–570 (2005).
21. G.T. Belz, G.M. Behrens, C.M. Smith, J.F. Miller, C. Jones, K. Lejon, C.G. Fathman, S.N. Mueller , K. Shortman, F.R. Carbone, and W.R. Heath. The CD8a⁺ dendritic cell is responsible for inducing peripheral self-tolerance to tissue-associated antigens. *J Exp Med* **196**:1099–1104 (2002).
22. J.M. den Haan, S.M. Lehar and M.J. Bevan. CD8⁺ but not CD8⁻ dendritic cells cross-prime cytotoxic T cells in vivo. *J Exp Med* **192**:1685–1696 (2000).
23. S. Hugues, E. Mougneau, W. Ferlin, D. Jeske, P. Hofman, D. Homann, L. Beaudoin, C. Schrike, M. Von Herrath, A. Lehuen, and N. Glaichenhaus. Tolerance to islet antigens and prevention from diabetes induced by limited apoptosis of pancreatic beta cells. *Immunity* **16**:169–181 (2002).
24. C.M. Smith, G.T. Belz, N.S. Wilson, J.A. Villadangos, K. Shortman, F.R. Carbone and W.R. Heath. Cutting edge: conventional CD8a⁺ dendritic cells are preferentially involved in CTL priming after footpad infection with herpes simplex virus-1. *J Immunol* **170**:4437–4440 (2003).
25. G.T. Belz, C.M. Smith, D. Eichner, K. Shortman, G. Karupiah and W.R. Heath. Cutting edge: conventional CD8a⁺ dendritic cells are generally involved in priming CTL immunity to viruses. *J Immunol* **172**:1996–2000 (2004).
26. R.S. Allan, C.M. Smith, G. Belz, A.L. van Lint, L.M. Wakim, W.R. Heath and F.R. Carbone. Epidermal viral immunity is induced by CD8a⁺ dendritic cells but not Langerhans cells. *Science* **301**:1925–1928 (2003).
27. G.T. Belz, N.S. Wilson, C.M. Smith, A. Mount, F.R. Carbone and W.R. Heath. Bone marrow-derived cells expand memory CD8⁺ T cells to viral infections of the lung and skin. *Eur J Immunol*. In press.
28. N. Shastri and F. Gonzalez. Endogenous generation and presentation of the ovalbumin peptide/Kb complex to T cells. *J Immunol* **150**:2724–2736 (1993).
29. P. Borrow, C.F. Evans and M.B. Oldstone. Virus-induced immunosuppression: immune system-mediated destruction of virus-infected dendritic cells results in generalized immune suppression. *J Virol* **69**:1059–1070 (1995).
30. B. Odermatt, M. Eppler, T.P. Leist, H. Hengartner and R.M. Zinkernagel. Virus-triggered acquired immunodeficiency by cytotoxic T-cell-dependent destruction of antigen-presenting cells and lymph follicle structure. *Proc Natl Acad Sci USA*. **88**:8252–8256 (1991).
31. N.V. Serbina, T.P. Salazar-Mather, C.A. Biron, W.A. Kuziel and E.G. Pamer. TNF/iNOS-producing dendritic cells mediate innate immune defense against bacterial infection. *Immunity* **19**:59–70 (2003).
32. M. Dalod, T.P. Salazar-Mather, L. Malmgaard, C. Lewis, C. Asselin-Paturel, F. Briere, G. Trinchieri and C.A. Biron. Interferon alpha/beta and interleukin 12 responses to viral infections: pathways regulating dendritic cell cytokine expression in vivo. *J Exp Med* **195**:517–528 (2002).
33. C. Asselin-Paturel, A. Boonstra, M. Dalod, I. Durand, N. Yessaad, C. Dezutter-Dambuyant, A. Vicari, A. O'Garra, C. Biron, F. Briere and G. Trinchieri. Mouse type I IFN-producing cells are immature APCs with plasmacytoid morphology. *Nat Immunol* **2**:1144–1150 (2001).

34. G.T. Belz, W.R. Heath and F.R. Carbone. The role of dendritic cell subsets in selection between tolerance and immunity. *Immunol Cell Biol* **80**:463–468 (2002).

35. C. Filippi, S. Hugues, J. Cazareth, V. Julia, N. Glaichenhaus and S. Ugolini. CD4$^+$ T cell polarization in mice is modulated by strain-specific major histocompatibility complex-independent differences within dendritic cells. *J Exp Med* **198**:201–209 (2003).

36. X. Zhao, E. Deak, K. Soderberg, M. Linehan, D. Spezzano, J. Zhu, D.M. Knipe and A. Iwasaki. Vaginal submucosal dendritic cells, but not Langerhans cells, induce protective Th1 responses to herpes simplex virus-2. *J Exp Med* **197**:153–162 (2003).

37. G.T. Belz, K. Shortman, M.J. Bevan and W.R. Heath. CD8a$^+$ dendritic cells selectively present MHC class I-restricted noncytolytic viral and intracellular bacterial antigens in vivo. *J Immunol* **175**:196–200 (2005).

38. N.V. Serbina, W. Kuziel, R. Flavell, S. Akira, B. Rollins and E.G. Pamer. Sequential MyD88-independent and -dependent activation of innate immune responses to intracellular bacterial infection. *Immunity* **19**:891–901 (2003).

39. M.P. Lemos, F. Esquivel, P. Scott and T.M. Laufer. MHC class II expression restricted to CD8alpha+ and CD11b+ dendritic cells is sufficient for control of Leishmania major. *J Exp Med* **199**:725–730 (2004).

40. S.R. Crowe, S.J. Turner, S.C. Miller, A.D. Roberts, R.A. Rappolo, P.C. Doherty, K.H. Ely and D.L. Woodland. Differential antigen presentation regulates the changing patterns of CD8$^+$ T cell immunodominance in primary and secondary influenza virus infections. *J Exp Med* **198**:399–410 (2003).

41. M.F. Bachmann, A. Gallimore, S. Linkert, V. Cerundolo, A. Lanzavecchia, M. Kopf and A. Viola. Developmental regulation of Lck targeting to the CD8 coreceptor controls signaling in naive and memory T cells. *J Exp Med* **189**:1521–1530 (1999).

42. E.M. Bertram, W. Dawicki, B. Sedgmen, J.L. Bramson, D.H. Lynch and T.H. Watts. A switch in costimulation from CD28 to 4-1BB during primary versus secondary CD8 T cell response to influenza in vivo. *J Immunol* **172**:981–988 (2004).

43. W. Dawicki and T.H. Watts. Expression and function of 4-1BB during CD4 versus CD8 T cell responses in vivo. *Eur J Immunol* **34**:743–751 (2004).

3

UNDERSTANDING THE ROLE OF INNATE IMMUNITY IN THE MECHANISM OF ACTION OF THE LIVE ATTENUATED YELLOW FEVER VACCINE 17D

Troy D. Querec and Bali Pulendran

1. INTRODUCTION: A HISTORICAL PERSPECTIVE

The live attenuated Yellow Fever Vaccine 17D [YF-17D] is one of the most effective vaccines available. During the 70 years since its development, the vaccine has been administered to more than 400 million people worldwide with minimal incident of severe side effects. Despite its efficacy, the immunological mechanisms that mediate its efficacy are poorly understood. Here we review the development of YF-17D in a historical context, and then present some emerging evidence which suggests that YF-17D activates multiple Toll-like receptors (TLRs) on dendritic cells (DCs) to elicit a broad spectrum of innate and adaptive immune responses. Interestingly, the resulting adaptive immune responses are characterized by a mixed T helper cell (Th)1/Th2 cytokine profile and antigen-specific CD8(+) T cells, and distinct TLRs appear to differentially control the Th1/Th2 balance. These data offer some new insights into the molecular mechanism of action of YF-17D, and highlight the potential of vaccination strategies that use combinations of different TLR ligands to stimulate polyvalent immune responses.

Max Theiler and his associates at Rockefeller University created the yellow fever vaccine 17D (YF-17D) empirically by serial passaging of the virus in tissue culture[1]. The Asibi strain of the yellow fever virus, which had been isolated in West Africa in 1927 and passaged in rhesus monkeys, served as the starting material for the vaccine development effort. Based on experience that the tropism of the virus could be altered by serial passage through a particular kind of

Department of Pathology and Emory Vaccine Center, 954 Gatewood Road, Atlanta, Georgia 30329, USA.

tissue, Theiler chose to passage the virus in various animal tissues to attenuate its virulence in humans[2]. However, to function as a vaccine, the virus had to maintain its ability to induce a strong immune response. Thus various passages of the virus were also checked for the ability to produce neutralizing antibodies and protect monkeys from challenge with a virulent strain of the virus. Several subculture strategies were tried including passage in mouse testicle and guinea pig tissue. Most substrains resulted in virus that either was still too lethal in monkeys or did not induce an effective immune response. The one exception was substrain 17D. The virus was passaged over 200 times first in mouse brain and mouse embryo tissue culture to reduce its viscerotropism, which can lead to hepatic and renal failure and hemorrhage, and next in chicken embryo and chicken embryo tissue culture with the brain and spinal cord removed to attenuate its neurotropism[2,3]. This work culminated in the 1937 *Journal of Experimental Medicine* paper in which YF-17D was first used in human volunteers, including the authors, and shown to induce neutralizing antibodies[4]. In 1945 the World Health Organization organized the establishment of two substrain seed lots of this vaccine: 17DD, which is used in South America, and 17D-204, which is used in the majority of the rest of the world[1]. Both substrains of the vaccine are currently manufactured with embryonated chicken eggs.

The success of YF-17D depends on a delicate balance between attenuating the virus so as to eliminate virulence without attenuation to such an extent that it loses its immunogenicity. Interestingly, there are very few potentially key mutations that are involved in attenuation of the vaccine. A comparison of YF-17D with wild-type Asibi reveals that there are only 68 nucleotide mutations out of over 10,860 nucleotides (~0.63%) that generate 32 amino acid differences[5]. The E protein appears to be the most heavily changed region of the entire genome, with 11 nucleotide and 8 amino acid differences. Because of the role the E protein plays in cell entry, some of these mutations must play a role in the tropism of the virus and thus affect its virulence. Based on their locations in the domains of the E protein, mutations 52 Gly→Arg and 200 Lys→Thr might affect the low pH endosomal conformational changes and 305 Ser→Phe and 380 Thr→Arg might affect cell attachment[5-9]. Also, studies have shown that by neural tissue passage YF-17D can be converted into a neurovirulent virus by mutations in the E protein[7,9-13]. Another interesting observation is that mutation of 3′-UTR is speculated to play a role in attenuation[14]. Four nucleotide differences occur in 3′-UTR, while none occur in 5′-UTR. Thermodynamically stable hairpin loops are present in the UTRs. Structural polymorphisms exist among the different substrains of YFV, and particular polymorphisms are associated with virulent strains versus attenuated vaccine strains. These structural differences in 3′-UTR presumably affect the transcription and translation of the virus by altering the recruitment of host and viral proteins.

Numerous clinical studies have been conducted with YF-17D to analyze viremia, cytokines, and acquired immune responses. Typically, viremia is detectable between 2 and 6 days post-vaccination as measured by RT-PCR in 30 to

100% of vaccinees depending on the study[15]. Plaque assays show that vaccine viremia averages 20 PFU/ml and does not exceed 100 PFU/ml[15], as opposed to 7 log10 with wild-type YF[16]. Plaque assays are a less sensitive method to measure YFV. Only a subset of cases positive by RT-PCR are also positive by plaque assay, and in those cases that are positive by plaque assay RT-PCR still detects viremia 1 day after no plaques are detected[15]. The inflammatory markers TNF, IL-1R antagonist, and IL-6 have been shown to peak in plasma at 2 days post-vaccination in response to the initial inoculum[17]. A second minor peak is found around day 7, presumably in response to the virus that has replicated in vivo and disseminated throughout the body. In contrast, IFN-alpha levels are not detectable until at least day 4 and peak around day 6, which is about a day after the peak in viremia[18]. The kinetics of the IFN-alpha response suggest that it is released into the blood possibly by circulating plasmacytoid DCs in response to viremia, as opposed to virus infecting tissues.

The strong antibody response to YF-17D has been well characterized. Neutralizing antibodies are believed to be the primary correlate of protection against challenge with wild-type YFV, and the appearance of neutralizing antibodies about 1 week post-vaccination correlates with the endpoint of vaccine viremia[15,18–20]. Individuals that receive a second vaccination of YF-17D do not have detectable viremia, presumably because of the presence of neutralizing antibodies[15]. An analysis of World War II veterans has shown that a single dose of the vaccine is capable of inducing detectable neutralizing antibody titers in over 80% of the vaccinees for more that 30 years[21,22]. In recent studies, seroconversion rates range between 90 and 100% of vaccinees 7 to 21 days after vaccination[23,24].

Both IgM and IgG antibodies are induced by primary vaccination. It is the IgG antibodies that are believed to be responsible for protection years after vaccination, because only IgG and not IgM antibodies are found after a secondary vaccination with YF-17D[19,25]. However, IgM titers can last for 6 months to a year, and one study has even tracked neutralizing IgM for several years[25]. The primary antibody targets for YFV immunity are the E protein and NS1[26,27]. Neutralization of E protein inhibits binding of the virus to the cell. Because NS1 is expressed on the surface of cells, antibody binding to this protein can promote antibody-dependent cell-mediated cytotoxicity (ADCC) and complement-mediated lysis to clear infected cells.

Though neutralizing antibody responses historically have been the focus of studies on protection against yellow fever, T cells do play an important role at least during the primary response to the vaccine. Only a few studies have investigated T cell responses to YF-17D. Based on general knowledge of antiviral T cell responses, it can be assumed that cytotoxic CD8 T cell responses control the infection by killing infected cells, and CD4 T cells mature into either Th1 or Th2 cells to promote stronger cytotoxic T cell and antibody responses. Human CD8 T cell epitopes to YF-17D have been found in E, NS1, NS2b, and NS3 proteins[28]. YF-17D-specific T cells can be measured by IFN-gamma ELISPOTs at 14 days post-vaccination and have been detected as far out as 19 months post-vaccination[28]. In C57Bl/6 mice, three T cell epitopes have been documented[29]. While the only detectable CD4 T cell response was to an I-Ab-restricted E protein epitope, responses to a dominant H-2Kb-restricted NS3 protein epitope and subdominant H-2Db-restricted E protein were measured in CD8 T cells.

2. UNDERSTANDING THE INNATE IMMUNE MECHANISM OF ACTION OF YF-17D

Despite YF-17D's efficacy in controlling yellow fever over the past several decades and its promise as a vaccine vector, a fundamental question has gone unanswered: How does YF-17D work so well? Recent advances in immunology suggest that the innate immune system is a critical determinant of the strength and quality of the adaptive immune response[30-34]. Within the innate immune system, dendritic cells (DCs) occupy a preeminent position, as they play critical roles in sensing microbial stimuli, and in initiating and modulating adaptive immune responses. DCs of an immature type are scattered throughout the body, particularly at the portals of pathogen entry, where they are equipped to sense microbial signatures through pathogen recognition receptors (PRRs), such as the Toll-like receptors (TLRs)[31,35,36]. TLRs constitute an evolutionarily conserved family of receptors involved in microbe recognition, of which 11 have been described in mammals. Different TLRs appear to recognize distinct microbial components. For example, lipopolysaccharides from *E. coli* are recognized by TLR 4[37], while certain lipopeptides — LPS from leptospira[38,39] and *P. gingivalis*[40,41] and the yeast cell wall zymosan[42] — are recognized by TLR 2. Furthermore, unmethylated DNA from bacteria and viruses are recognized by TLR 9[43-46], single-stranded RNA is recognized by TLR 7/8[47-49], and double-stranded RNA is recognized by TLR 3[50-52].

Upon recognition of a pathogen via TLRs, the immature DCs at the site of pathogen entry undergo a maturation process, during which they exit the site and migrate to the T cell-rich areas of the neighboring lymph nodes, whereas the mature DCs present their acquired microbial antigens and stimulate antigen-specific T cells, thus initiating adaptive immunity and immune memory.

Given the pivotal role of TLRs and DCs in initiating and tuning the adaptive immune response, there is at present much interest in exploiting these in the development of novel vaccines. However, the converse question of whether some of our best, empirically derived vaccines actually work by stimulating TLRs is largely unexplored. In this context, in our recent study we sought to determine whether YF-17D triggered activation of innate immunity and DCs via TLRs, and if so what effect this had on the quality of the innate and adaptive immune responses.

With their key function in immune modulation, DCs likely have an important role to play in YF-17D vaccination. Human monoycte-derived DCs can be infected with YF-17D, similar to other flaviviruses such as dengue and West

Nile[53-60]. Dengue and West Nile viruses have been shown to replicate in Langerhans cells in the skin. Recently yellow fever vaccines have been shown to replicate in DCs[53,59,60]. This replication in DCs could enhance the immune response by facilitating transport of viral antigens to the lymph nodes, increase the concentration of antigens available for presentation, and stimulate a stronger danger signal. As a further enhancement of DC induction of acquired immune responses, flaviviruses cause viral-mediated upregulation of MHC class I[61-63].

Our recent findings suggest that YF-17D activates multiple Toll-like receptors (TLRs), on distinct subsets of dendritic cells (DCs), to elicit a broad spectrum of innate and adaptive immune responses[59]. Specifically, YF-17D activates multiple DC subsets via TLRs 2, 7, 8, and 9 to elicit potent proinflammatory cytokines, as well as IFNα. Interestingly, the resulting adaptive immune responses are characterized by a mixed Th1/Th2 cytokine profile and antigen-specific CD8+ T cells. Furthermore, distinct TLRs appear to differentially control the Th1/Th2 balance: thus, while MyD88-deficient mice show a profound impairment of Th1 cytokines, TLR2-deficient mice show greatly enhanced Th1 and Tc1 responses to YF-17D. Two intriguing features of the data deserve discussion. First, it was surprising that deficiency of a single TLR resulted in a marked impairment of IL-12p40 and IL-6, since it might have been expected that other TLRs would compensate for this deficiency. One explanation is that YF-17D contains suboptimal amounts of individual TLR ligands, which by themselves stimulate suboptimal responses but can signal in concert to create a synergistic effect. This hypothesis is consistent with the recent finding that selected TLR combinations can indeed synergistically trigger a Th1 polarizing program in human DCs[64,65]. A second intriguing feature is that signaling via distinct TLRs appears to exert opposing influences on the Th1/Th2 balance. Thus, eliminating signaling via MyD88 impairs Th1 responses, while eliminating TLR2 signaling enhances Th1 responses. This is consistent with emerging data that distinct TLRs may differentially affect the Th1/Th2 balance[65-71]. Therefore, triggering multiple TLRs might facilitate both immune synergy and generation of polyvalent immune responses.

The question of which viral components act through which innate immune receptors is one of significant importance. Since live viral vaccines contain not only viral proteins and nucleic acids but also cellular contaminants derived from the cell lines in which the vaccines were grown, the latter may play important roles in the induction and modulation of vaccine induced immune responses. Thus, the chicken embryos and cell lines in which YF-17D are grown are potential sources of these dirty little secrets or danger signals, and it is thus likely that these factors and other nonviral components of the YF-17D vaccine, such as heat shock proteins and uric acid, contribute to the induction of an immune response. However, whatever the source of the molecules that stimulate a given TLR or PRR, further work is clearly required to determine the specific combinations of PRRs (the correct "code" of innate immune buttons) that promote a particular facet of the immune response, such as Th1/Th2 balance, neutralizing antibody responses, long-lived antibody-producing cells that home to mucosal tissues and continuously secrete high-affinity neutralizing antibody, and long-lived memory T cells.

3. CONCLUDING REMARKS

In conclusion, recent findings enhance our understanding of the immunological mechanisms of action of YF-17D and suggest that vaccine adjuvants that contain multiple TLRs might be beneficial in stimulating a broad spectrum of innate and adaptive responses. Therefore, determining the TLRs (and other non-TLR pathogen recognition receptors) through which YF-17D and other effective vaccines signal could be used to deconstruct the immunological basis for the efficacy of such "successful" vaccines. It is likely that many of these vaccines, which consist of live attenuated viruses or bacteria, may well signal through multiple TLRs, as well as other non-TLR receptors such as RIG-1[72]. Conversely, vaccines that stimulate suboptimal responses upon a single injection, but which require multiple boosts to generate robust protective responses, may fail to trigger TLRs, or may even inhibit TLR signaling. Thus, future vaccine strategies aimed at incorporating multiple TLR ligands and antigens into delivery systems such as nanoparticles, which can target specific DC subsets in vivo, are likely to be of great benefit in conferring the degree of immunogenicity and protection mediated by our best vaccines — such as the yellow fever vaccine YF-17D.

4. ACKNOWLEDGMENTS

This work was generously supported by grants from the National Institutes of Health, nos. U19 AI057266, R01 AI048638, R01 AI056499, R01 DK057665, R21 AI056957, and U54 AI057157. The authors have no conflicting financial interests.

5. REFERENCES

1. T.P. Monath. Milestones the conquest of yellow fever. In *Microbe hunters: then and now*. Ed. H. Koprowski and M.B.A. Oldstone. Lansing, MI: Medi-Ed Press (1996).
2. M. Theiler and H.H. Smith. The effect of prolonged cultivation in vitro upon the pathogenicity of yellow fever virus. *J Exp Med* **65**(6):767–786 (1937).
3. M. Theiler and H.H. Smith. The use of yellow fever virus modified by in vitro cultivation for human immunization. *J Exp Med* **65**:787–800 (1937). *Rev Med Virol* **10**(1):6–16; discussion 13–15 (2000).
4. M. Theiler and H.H. Smith. The use of yellow fever virus modified by in vitro cultivation for human immunization. *J Exp Med* **65**(6):787–800 (1937).

5. C.S. Hahn, J.M. Dalrymple, J.H. Strauss and C.M. Rice. Comparison of the virulent asibi strain of yellow fever virus with the 17d vaccine strain derived from it. *Proc Natl Acad Sci USA* **84**(7):2019–2023 (1987).

6. S.L. Allison, J. Schalich, K. Stiasny, C.W. Mandl and F.X. Heinz. Mutational evidence for an internal fusion peptide in flavivirus envelope protein e. *J Virol* **75**(9):4268–4275 (2001).

7. K.D. Ryman, H. Xie, T.N. Ledger, G.A. Campbell and A.D. Barrett. Antigenic variants of yellow fever virus with an altered neurovirulence phenotype in mice. *Virology* **230**(2):376–380 (1997).

8. L. Vlaycheva, M. Nickells, D.A. Droll and T.J. Chambers. Yellow fever 17d virus: Pseudo–revertant suppression of defective virus penetration and spread by mutations in domains ii and iii of the e protein. *Virology* **327**(1):41–49 (2004).

9. F. Guirakhoo, Z. Zhang, G. Myers, B.W. Johnson, K. Pugachev, R. Nichols, N. Brown, I. Levenbook, K. Draper, S. Cyrek, J. Lang, C. Fournier, B. Barrere, S. Delagrave and T.P. Monath. A single amino acid substitution in the envelope protein of chimeric yellow fever-dengue 1 vaccine virus reduces neurovirulence for suckling mice and viremia/viscerotropism for monkeys. *J. Virol.* **78**(18):9998–10008 (2004).

10. K.D. Ryman, T.N. Ledger, G.A. Campbell, S.J. Watowich and A.D. Barrett. Mutation in a 17d-204 vaccine substrain-specific envelope protein epitope alters the pathogenesis of yellow fever virus in mice. *Virology* **244**(1):59–65 (1998).

11. T.P. Monath, J. Arroyo, I. Levenbook, Z.X. Zhang, J. Catalan, K. Draper and F. Guirakhoo. Single mutation in the flavivirus envelope protein hinge region increases neurovirulence for mice and monkeys but decreases viscerotropism for monkeys: relevance to development and safety testing of live attenuated vaccines. *J Virol* **76**(4):1932–1943 (2002).

12. T.J. Chambers and M. Nickells. Neuroadapted yellow fever virus 17d: Genetic and biological characterization of a highly mouse-neurovirulent virus and its infectious molecular clone. *J Virol* **75**(22):10912–10922 (2001).

13. J.J. Schlesinger, S. Chapman, A. Nestorowicz, C.M. Rice, T.E. Ginocchio and T.J. Chambers. Replication of yellow fever virus in the mouse central nervous system: comparison of neuroadapted and non-neuroadapted virus and partial sequence analysis of the neuroadapted strain. *J Gen Virol* **77**(Pt 6):1277–1285 (1996).

14. V. Proutski, M.W. Gaunt, E.A. Gould and E.C. Holmes. Secondary structure of the 3'-untranslated region of yellow fever virus: implications for virulence, attenuation and vaccine development. *J Gen Virol* **78**(Pt 7):1543–1549 (1997).

15. B. Reinhardt, R. Jaspert, M. Niedrig, C. Kostner and J. L'Age-Stehr. Development of viremia and humoral and cellular parameters of immune activation after vaccination with yellow fever virus strain 17d: a model of human flavivirus infection. *J Med Virol* **56**(2):159–167 (1998).

16. F.N. Macnamara. A clinico-pathological study of yellow fever in nigeria. *West Afr Med J* **6**(4):137–146 (1957).

17. U.T. Hacker, T. Jelinek, S. Erhardt, A. Eigler, G. Hartmann, H.D. Nothdurft and S. Endres. In vivo synthesis of tumor necrosis factor-alpha in healthy humans after live yellow fever vaccination. *J Infect Dis* **177**(3):774–778 (1998).

18. E.F. Wheelock and W.A. Sibley. Circulating virus, interferon and antibody after vaccination with the 17-d strain of yellow-fever virus. *N Engl J Med* **273**:194–198 (1965).

19. V. Bonnevie-Nielsen, I. Heron, T.P. Monath and C.H. Calisher. Lymphocytic 2',5'-oligoadenylate synthetase activity increases prior to the appearance of neutralizing antibodies and immunoglobulin m and immunoglobulin g antibodies after primary and secondary immunization with yellow fever vaccine. *Clin Diagn Lab Immunol* **2**(3):302–306 (1995).

20. J. Lang, J. Zuckerman, P. Clarke, P. Barrett, C. Kirkpatrick and C. Blondeau. Comparison of the immunogenicity and safety of two 17d yellow fever vaccines. *Am J Trop Med Hyg* **60**(6):1045–1050 (1999).

21. M. Niedrig, M. Lademann, P. Emmerich and M. Lafrenz. Assessment of igg antibodies against yellow fever virus after vaccination with 17d by different assays: Neutralization test, haemagglutination inhibition test, immunofluorescence assay and elisa. *Trop Med Int Health* **4**(12):867–871 (1999).

22. J.D. Poland, C.H. Calisher, T.P. Monath, W.G. Downs and K. Murphy. Persistence of neutralizing antibody 30-35 years after immunization with 17d yellow fever vaccine. *Bull World Health Organ* **59**(6):895–900 (1981).

23. V.E. Belmusto-Worn, J.L. Sanchez, K. McCarthy, R. Nichols, C.T. Bautista, A.J. Magill, G. Pastor-Cauna, C. Echevarria, V.A. Laguna-Torres, B.K. Samame, M.E. Baldeon, J.P. Burans, J.G. Olson, P. Bedford, S. Kitchener and T.P. Monath. Randomized, double-blind, phase iii, pivotal field trial of the comparative immunogenicity, safety, and tolerability of two yellow fever 17d vaccines (arilvax and yf-vax) in healthy infants and children in Peru. *Am J Trop Med Hyg* **72**(2):189–197 (2005).

24. T.P. Monath, R. Nichols, W.T. Archambault, L. Moore, R. Marchesani, J. Tian, R.E. Shope, N. Thomas, R. Schrader, D. Furby and P. Bedford. Comparative safety and immunogenicity of two yellow fever 17d vaccines (arilvax and yf-vax) in a phase iii multicenter, double-blind clinical trial. *Am J Trop Med Hyg* **66**(5):533–541 (2002).

25. T.P. Monath. Neutralizing antibody responses in the major immunoglobulin classes to yellow fever 17d vaccination of humans. *Am J Epidemiol* **93**(2):122–129 (1971).

26. S. Daffis, R.E. Kontermann, J. Korimbocus, H. Zeller, H.D. Klenk and J. Ter Meulen. Antibody responses against wild-type yellow fever virus and the 17d vaccine strain: characterization with human monoclonal antibody fragments and neutralization escape variants. *Virology* **337**(2):262–272 (2005).

27. J.J. Schlesinger, M.W. Brandriss, J.R. Putnak and E.E. Walsh. Cell surface expression of yellow fever virus non-structural glycoprotein ns1: consequences of interaction with antibody. *J Gen Virol* **71**(Pt 3):593–599 (1990).

28. M.D. Co, M. Terajima, J. Cruz, F.A. Ennis and A.L. Rothman. Human cytotoxic t lymphocyte responses to live attenuated 17d yellow fever vaccine: identification of hla-b35-restricted ctl epitopes on nonstructural proteins ns1, ns2b, ns3, and the structural protein e. *Virology* **293**(1):151–163 (2002).

29. R.G. van der Most, L.E. Harrington, V. Giuggio, P.L. Mahar and R. Ahmed. Yellow fever virus 17d envelope and ns3 proteins are major targets of the antiviral t cell response in mice. *Virology* **296**(1):117–124 (2002).

30. B. Beutler. Inferences, questions and possibilities in toll-like receptor signalling. *Nature* **430**(6996):257–263 (2004).

31. A. Iwasaki and R. Medzhitov. Toll-like receptor control of the adaptive immune responses. *Nat Immunol* **5**(10):987–995 (2004).

32. K. Takeda, T. Kaisho and S. Akira. Toll-like receptors. *Annu Rev Immunol* **21**:335–376 (2003).

33. T. Kawai and S. Akira. Pathogen recognition with toll-like receptors. *Curr Opin Immunol* **17**(4):338–344 (2005).
34. R.N. Germain. An innately interesting decade of research in immunology. *Nat Med* **10**(12):1307–1320 (2004).
35. K. Shortman and Y.J. Liu. Mouse and human dendritic cell subtypes. *Nat Rev Immunol* **2**(3):151–161 (2002).
36. B. Pulendran. Variegation of the immune response with dendritic cells and pathogen recognition receptors. *J Immunol* **174**(5):2457–2465 (2005).
37. A. Poltorak, X. He, I. Smirnova, M.Y. Liu, C. Van Huffel, X. Du, D. Birdwell, E. Alejos, M. Silva, C. Galanos, M. Freudenberg, P. Ricciardi-Castagnoli, B. Layton and B. Beutler. Defective lps signaling in c3h/hej and c57bl/10sccr mice: mutations in tlr4 gene. *Science* **282**(5396):2085–2088 (1998).
38. O. Takeuchi, K. Hoshino, T. Kawai, H. Sanjo, H. Takada, T. Ogawa, K. Takeda and S. Akira. Differential roles of tlr2 and tlr4 in recognition of gram-negative and gram-positive bacterial cell wall components. *Immunity* **11**(4):443–451 (1999).
39. C. Werts, R.I. Tapping, J.C. Mathison, T.H. Chuang, V. Kravchenko, I. Saint Girons, D.A. Haake, P.J. Godowski, F. Hayashi, A. Ozinsky, D.M. Underhill, C.J. Kirschning, H. Wagner, A. Aderem, P.S. Tobias and R.J. Ulevitch. Leptospiral lipopolysaccharide activates cells through a tlr2-dependent mechanism. *Nat Immunol* **2**(4):346–352 (2001).
40. M. Hirschfeld, J.J. Weis, V. Toshchakov, C.A. Salkowski, M.J. Cody, D.C. Ward, N. Qureshi, S.M. Michalek and S.N. Vogel. Signaling by toll-like receptor 2 and 4 agonists results in differential gene expression in murine macrophages. *Infect Immunol* **69**(3):1477–1482 (2001).
41. H. Hiramine, K. Watanabe, N. Hamada and T. Umemoto. Porphyromonas gingivalis 67-kda fimbriae induced cytokine production and osteoclast differentiation utilizing tlr2. *FEMS Microbiol Lett* **229**(1):49–55 (2003).
42. D.M. Underhill, A. Ozinsky, A.M. Hajjar, A. Stevens, C.B. Wilson, M. Bassetti and A. Aderem. The toll-like receptor 2 is recruited to macrophage phagosomes and discriminates between pathogens. *Nature* **401**(6755):811–815 (1999).
43. A.M. Krieg. Cpg motifs in bacterial DNA and their immune effects. *Annu Rev Immunol* **20**:709–760 (2002).
44. A. Krug, A.R. French, W. Barchet, J.A. Fischer, A. Dzionek, J.T. Pingel, M.M. Orihuela, S. Akira, W.M. Yokoyama and M. Colonna. Tlr9-dependent recognition of mcmv by ipc and dc generates coordinated cytokine responses that activate antiviral nk cell function. *Immunity* **21**(1):107–119 (2004).
45. J. Lund, A. Sato, S. Akira, R. Medzhitov and A. Iwasaki. Toll-like receptor 9-mediated recognition of herpes simplex virus-2 by plasmacytoid dendritic cells. *J Exp Med* **198**(3):513–520 (2003).
46. F. Takeshita, C.A. Leifer, I. Gursel, K.J. Ishii, S. Takeshita, M. Gursel and D.M. Klinman. Cutting edge: Role of toll-like receptor 9 in cpg DNA-induced activation of human cells. *J Immunol* **167**(7):3555–3558 (2001).
47. S.S. Diebold, T. Kaisho, H. Hemmi, S. Akira and C. Reis e Sousa. Innate antiviral responses by means of tlr7-mediated recognition of single-stranded rna. *Science* **303**(5663):1529–1531 (2004).
48. F. Heil, H. Hemmi, H. Hochrein, F. Ampenberger, C. Kirschning, S. Akira, G. Lipford, H. Wagner and S. Bauer. Species-specific recognition of single-stranded rna via toll-like receptor 7 and 8. *Science* **303**(5663):1526–1529 (2004).

49. J.M. Lund, L. Alexopoulou, A. Sato, M. Karow, N.C. Adams, N.W. Gale, A. Iwasaki and R.A. Flavell. Recognition of single-stranded rna viruses by toll-like receptor 7. *Proc Natl Acad Sci USA* **101**(15):5598–5603 (2004).
50. L. Alexopoulou, A.C. Holt, R. Medzhitov and R.A. Flavell. Recognition of double-stranded rna and activation of nf-kappab by toll-like receptor 3. *Nature* **413**(6857):732–738 (2001).
51. L. Guillot, R. Le Goffic, S. Bloch, N. Escriou, S. Akira, M. Chignard and M. Si-Tahar. Involvement of toll-like receptor 3 in the immune response of lung epithelial cells to double-stranded rna and influenza a virus. *J Biol Chem* **280**(7):5571–5580 (2005).
52. K. Tabeta, P. Georgel, E. Janssen, X. Du, K. Hoebe, K. Crozat, S. Mudd, L. Shamel, S. Sovath, J. Goode, L. Alexopoulou, R.A. Flavell and B. Beutler. Toll-like receptors 9 and 3 as essential components of innate immune defense against mouse cytomegalovirus infection. *Proc Natl Acad Sci USA* **101**(10):3516–3521 (2004).
53. S. Brandler, N. Brown, T.H. Ermak, F. Mitchell, M. Parsons, Z. Zhang, J. Lang, T.P. Monath and F. Guirakhoo. Replication of chimeric yellow fever virus-dengue serotype 1-4 virus vaccine strains in dendritic and hepatic cells. *Am J Trop Med Hyg* **72**(1):74–81 (2005).
54. S.J. Wu, G. Grouard-Vogel, W. Sun, J.R. Mascola, E. Brachtel, R. Putvatana, M.K. Louder, L. Filgueira, M.A. Marovich, H.K. Wong, A. Blauvelt, G.S. Murphy, M.L. Robb, B.L. Innes, D.L. Birx, C.G. Hayes and S.S. Frankel. Human skin langerhans cells are targets of dengue virus infection. *Nat Med* **6**(7):816–820 (2000).
55. L.J. Ho, J.J. Wang, M.F. Shaio, C.L. Kao, D.M. Chang, S.W. Han and J.H. Lai. Infection of human dendritic cells by dengue virus causes cell maturation and cytokine production. *J Immunol* **166**(3):1499–1506 (2001).
56. S.N. Byrne, G.M. Halliday, L.J. Johnston and N.J. King. Interleukin-1beta but not tumor necrosis factor is involved in west nile virus-induced langerhans cell migration from the skin in c57bl/6 mice. *J Invest Dermatol* **117**(3):702–709 (2001).
57. V.B. Pisarev, E.O. Shishkina, V.F. Larichev and N.V. Grigor'eva. Morphofunctional characteristics of antigen-presenting cells in lymph node in mice with experimental west nile fever. *Bull Exp Biol Med* **135**(3):293–295 (2003).
58. L.J. Johnston, G.M. Halliday and N.J. King. Langerhans cells migrate to local lymph nodes following cutaneous infection with an arbovirus. *J Invest Dermatol* **114**(3):560–568 (2000).
59. T. Querec, S. Bennouna, S. Alkan, Y. Laouar, K. Gorden, R. Flavell, S. Akira, R. Ahmed and B. Pulendran. Yellow fever vaccine yf-17d activates multiple dendritic cell subsets via tlr2, 7, 8, and 9 to stimulate polyvalent immunity. *J Exp Med* **203**(2):413–424 (2006).
60. G. Barba-Spaeth, R.S. Longman, M.L. Albert and C.M. Rice. Live attenuated yellow fever 17d infects human dcs and allows for presentation of endogenous and recombinant t cell epitopes. *J Exp Med* **202**(9):1179–1184 (Epub 2005 Oct 1131) (2005).
61. M. Lobigs, A. Mullbacher and M. Regner. Mhc class i up-regulation by flaviviruses: immune interaction with unknown advantage to host or pathogen. *Immunol Cell Biol* **81**(3):217–223 (2003).
62. F. Momburg, A. Mullbacher and M. Lobigs. Modulation of transporter associated with antigen processing (tap)-mediated peptide import into the endoplasmic reticulum by flavivirus infection. *J Virol* **75**(12):5663–5671 (2001).

63. A. Mullbacher and M. Lobigs. Up-regulation of mhc class i by flavivirus-induced peptide translocation into the endoplasmic reticulum. *Immunity* **3**(2):207–214 (1995).

64. M.F. Roelofs, L.A. Joosten, S. Abdollahi-Roodsaz, A.W. Van Lieshout, T. Sprong, F.H. van den Hoogen, W.B. Van Den Berg and T.R. Radstake. The expression of toll-like receptors 3 and 7 in rheumatoid arthritis synovium is increased and costimulation of toll-like receptors 3, 4, and 7/8 results in synergistic cytokine production by dendritic cells. *Arthr Rheum.* **52**(8):2313–2322 (2005).

65. G. Napolitani, A. Rinaldi, F. Bertoni, F. Sallusto and A. Lanzavecchia. Selected toll-like receptor agonist combinations synergistically trigger a t helper type 1-polarizing program in dendritic cells. *Nat Immunol* **6**(8):769–776 (2005).

66. B. Pulendran, P. Kumar, C.W. Cutler, M. Mohamadzadeh, T. Van Dyke and J. Banchereau. Lipopolysaccharides from distinct pathogens induce different classes of immune responses in vivo. *J Immunol* **167**(9):5067–5076 (2001).

67. S. Dillon, A. Agrawal, T. Van Dyke, G. Landreth, L. McCauley, A. Koh, C. Maliszewski, S. Akira and B. Pulendran. A toll-like receptor 2 ligand stimulates th2 responses in vivo, via induction of extracellular signal-regulated kinase mitogen-activated protein kinase and c-fos in dendritic cells. *J Immunol* **172**(8):4733–4743 (2004).

68. M.G. Netea, R. Sutmuller, C. Hermann, C.A. Van der Graaf, J.W. Van der Meer, J.H. van Krieken, T. Hartung, G. Adema and B.J. Kullberg. Toll-like receptor 2 suppresses immunity against candida albicans through induction of il-10 and regulatory t cells. *J Immunol* **172**(6):3712–3718 (2004).

69. A. Sing, D. Rost, N. Tvardovskaia, A. Roggenkamp, A. Wiedemann, C.J. Kirschning, M. Aepfelbacher and J. Heesemann. Yersinia v-antigen exploits toll-like receptor 2 and cd14 for interleukin 10-mediated immunosuppression. *J Exp Med* **196**(8):1017–1024 (2002).

70. V. Redecke, H. Hacker, S.K. Datta, A. Fermin, P.M. Pitha, D.H. Broide and E. Raz. Cutting edge: activation of toll-like receptor 2 induces a th2 immune response and promotes experimental asthma. *J Immunol* **172**(5):2739–2743 (2004).

71. S. Agrawal, A. Agrawal, B. Doughty, A. Gerwitz, J. Blenis, T. Van Dyke and B. Pulendran. Cutting edge: different toll-like receptor agonists instruct dendritic cells to induce distinct th responses via differential modulation of extracellular signal-regulated kinase-mitogen-activated protein kinase and c-fos. *J Immunol* **171**(10):4984–4989 (2003).

72. H. Kato, S. Sato, M. Yoneyama, M. Yamamoto, S. Uematsu, K. Matsui, T. Tsujimura, K. Takeda, T. Fujita, O. Takeuchi and S. Akira. Cell type-specific involvement of rig-i in antiviral response. *Immunity* **23**(1):19–28 (2005).

4

THE FUNCTION OF LOCAL LYMPHOID TISSUES IN PULMONARY IMMUNE RESPONSES

Juan Moyron-Quiroz, Javier Rangel-Moreno,
Damian M. Carragher, and Troy D. Randall*

1. LYMPH NODE STRUCTURE AND DEVELOPMENT

Primary adaptive immune responses are initiated in secondary lymphoid organs, such as spleen, lymph nodes, and Peyer's patches. These lymphoid organs recruit naive lymphocytes[1] as well as activated antigen-presenting cells (APCs)[2], and facilitate lymphocyte activation, expansion, and differentiation. For example, infection of the lung with influenza virus leads to activation of pulmonary dendritic cells, which engulf local antigens and traffic to the draining mediastinal lymph node (MLN)[3], where they home to the T cell area surrounding the high endothelial venules (HEVs) (Figure 1). Naive B and T cells are constantly recruited into the lymph node via these HEVs and rapidly become activated as they encounter cognate antigen on APCs. Activated lymphocytes subsequently expand and differentiate into effector cells. For T cells, this differentiation primarily occurs in the T cell zone. In contrast, B cells rapidly expand and are selected for high-affinity variants in the germinal centers (GCs) that develop on the border between the T cell area and the B cell follicle. As the immune response progresses, effector B and T cells leave the lymph node via the efferent lymphatics, which drain into the blood via the thoracic duct. Once in the blood, activated effector cells recirculate to sites of inflammation, including the original site of infection in the lung, and use their effector functions to combat infection. An important point of this model is that, while infection occurs locally in non-lymphoid organs, primary immune responses are initiated centrally in secondary lymphoid organs. This scheme is outlined in Figure 1.

---

*From the Trudeau Institute, Saranac Lake, New York 12983, USA.

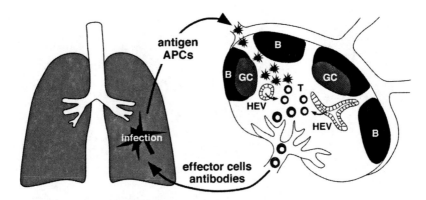

Figure 1. Secondary lymphoid organs acquire antigen and activated APCs from regional tissues and initiate primary immune responses. APCs enter the lymph node via afferent lymphatics and home to the T cell zone. Naive lymphocytes enter the lymph node via the HEVs and encounter antigen displayed on APCs. Effector lymphocytes leave the lymph node via efferent lymphatics.

Although there are similarities in the architecture of all secondary lymphoid organs, their structure differs depending on their location and function[4]. For example, the spleen samples antigens directly from the blood[5], while regional lymph nodes sample antigens from afferent lymphatics that drain regional tissues[4]. In contrast, mucosal lymphoid tissues, such as Peyer's patches and Nasal Associated Lymphoid Tissue (NALT) have no afferent lymphatics and sample antigens directly across the mucosal epithelium[6]. Therefore, each secondary lymphoid organ is specialized to acquire antigen from the regional tissues that it drains and to generate immune responses appropriate to those antigens.

Lymph node and Peyer's patch development occurs between day E11 of gestation and birth[7]. The development of these organs is dependent on the lymphotoxin (LT) signaling pathway and mice lacking LTα[8], LTβ[9], LTβR[10], or NFκB-inducing kinase[11] fail to develop most lymph nodes and Peyer's patches. Embryonic lymph node development is initiated by Lymphoid Tissue inducer cells (LTi cells), which provide lymphotoxin to local LTβR-expressing mesenchymal cells at sites of future lymph node development[12]. In turn, signals from the LTβR trigger mesenchymal cells to differentiate into mature stromal cells that express the chemokines and adhesion molecules necessary to recruit and maintain lymphocytes in mature lymph nodes[12]. In the absence of these chemokines, particularly CXCL13[13,14], LTi cells fail to migrate to sites of future lymph node formation and lymph nodes do not develop. Once formed, the architecture of secondary lymphoid organs is maintained by a positive feedback loop between lymphotoxin and the homeostatic chemokines[14]. In fact, lymphotoxin-dependent CXCL13 expression is required for maintenance of B cell follicles, formation of germinal centers, and differentiation of follicular dendritic cells[14]. In contrast, CCL21 is expressed primarily on HEVs and promotes rolling arrest and migration across the vascular endothelium[15], while CCL19 expression on stromal cells is important for recruitment of T cells and activated APCs into the T cell area[16]. Like the expression of CXCL13, the expression of CCL19 and CCL21 is dependent on lymphotoxin[17]. Thus, lymphotoxin and homeostatic chemokines are essential for both the development of secondary lymphoid organs and for maintenance of proper lymphoid architecture.

2. ROLE OF LOCAL LYMPHOID ORGANS IN PULMONARY IMMUNITY

Despite the important role of secondary lymphoid organs in initiating primary immune responses, there are some indications that naive lymphocytes can be primed locally in non-lymphoid tissues. For example, adoptively transferred naive OTII TCR transgenic CD4 T cells can be primed by intranasal challenge with OVA in $LT\alpha^{-/-}$ hosts[18]. Since $LT\alpha^{-/-}$ mice lack all lymph nodes and Peyer's patches, these data suggest that the transferred T cells are primed directly in the lung tissue by pulmonary DCs. We also used $LT\alpha^{-/-}$ mice in our initial studies to address whether conventional secondary lymphoid organs were required for immune responses to influenza[19]. As expected based on studies showing that $LT\alpha^{-/-}$ mice are unable to generate primary immune responses to a variety of infectious agents[20-23], we found that $LT\alpha^{-/-}$ mice are more susceptible than normal mice to influenza. However, even though the onset of immunity is slightly delayed, $LT\alpha^{-/-}$ mice make robust immune responses to influenza[19]. For example, although the initial CD8 T cell response is slightly delayed, normal numbers of influenza-specific CD8 T cells are eventually generated that have normal cytotoxic activity and make normal levels of IFNγ upon restimulation. The B cell response is also slightly delayed, but ultimately produces above-normal levels of influenza-specific IgM and slightly below normal levels of influenza-specific IgG. Together, the T and B cell responses of $LT\alpha^{-/-}$ mice clear influenza virus from the lungs with only a slight delay[19]. Thus, primary immune responses to influenza are intact in $LT\alpha^{-/-}$ mice, despite their lack of lymph nodes.

2.1. Structure and Function of Nasal-Associated Lymphoid Tissue (NALT)

Given that $LT\alpha^{-/-}$ mice are also able to make an immune response to another respiratory virus, γHerpesvirus-68[24], we hypothesized that there is something unique about the respiratory tract that allows primary immune responses to occur in the absence of lymph nodes. One possibility is that some secondary lymphoid organs remain in the respiratory tract of $LT\alpha^{-/-}$ mice. Since the loss of all conventional lymph nodes is extensively documented in $LT\alpha^{-/-}$ mice[8], we examined whether the development of NALT is disrupted by the loss of LTα. Surprisingly, we showed that, unlike all other lymph nodes and Peyer's patches,

NALT develops independently of LTα signaling[25]. However, despite the ability of NALT to bypass LTα for its development, the structure and function of NALT is severely compromised in the absence of LTα[25,26]. The NALT of LTα[-/-] mice is lymphopenic, lacks B cell follicles, follicular dendritic cells and HEVs, and is unable to initiate B and T cell responses to intranasal antigens. However, both the structure and immune function of NALT can be restored by reconstitution of LTα[-/-] mice with normal bone marrow[25]. Since conventional lymph nodes and Peyer's patches are not restored by reconstitution with normal cells[7], this suggests either that the basic scaffolding of NALT is formed independently of LTα, or that the development of NALT is not confined to an early developmental window. In either case, the development of NALT is unlike that of other secondary lymphoid organs.

The severely compromised structure and function of NALT in LTα[-/-] mice also suggests that LTα may control the proper expression of homeostatic chemokines in NALT as it does in the spleen. We investigated this possibility and found that, indeed, LTα is required for the expression of CXCL13, CCL19, and CCL21 in NALT and that the loss of these chemokines is responsible for the disrupted architecture and impaired function of NALT in LTα[-/-] mice[26]. For example, the architecture of NALT is severely disrupted and both B cell and T cell responses are impaired in LTα[-/-] mice. In contrast, only B cell responses are impaired in CXCL13[-/-] mice and only T cell responses are impaired in plt/plt mice, which lack both CCL19 and CCL21[26]. Thus, the structural and functional defects in LTα[-/-] NALT can be directly traced to impaired chemokine expression. These data are consistent with the severely reduced expression of CXCL13, CCL19, and CCL21[17], the compromised lymphoid architecture[27], and impaired function[28] of spleens in LTα[-/-] mice.

2.2. Pulmonary Immune Responses in the Absence of Secondary Lymphoid Organs

Although it is clear that NALT can develop in the absence of LTα, the inability of NALT to support B and T cell responses in LTα[-/-] mice suggest that immune responses to influenza can be initiated in a location other than NALT. To test this possibility, we next examined whether immune responses are intact in mice that lacked secondary lymphoid organs, but express LTα[29]. In these experiments, we splenectomized and irradiated LTα[-/-] mice and reconstituted them with normal bone marrow. These mice, which lack spleen, all lymph nodes, and Peyer's patches, are referred to as Spleen, Lymph node, and Peyer's patch-deficient mice (SLP mice). As controls, we reconstituted C57BL/6 mice with C57BL/6 bone marrow to generate WT mice (Figure 2A). After reconstitution, these mice were infected with influenza and assayed for T and B cell responses. As shown in Figure 2B, CD8 cells specific for influenza nucleoprotein$_{366-374}$ (NP) are easily identified in the lungs of WT animals on day 9, the peak of the CD8 T cell response to influenza in normal mice[19]. Influenza-specific CD8 cells are also detected in the lungs of SLP mice, although at a lower frequency than in WT mice (Figure 2B). To determine whether influenza-specific CD8 T cells ever accumulated to normal levels in SLP mice, we followed the kinetics of the CD8 T cell response. As expected, the number of influenza-specific CD8 T cells peaked in the lungs of WT mice on day 9 and then declined (Figure 2C). Despite the delayed appearance of influenza-specific CD8 T cells in the lungs of SLP mice, the number of influenza-specific CD8 T cells in the lungs of SLP mice reached WT levels by day 14 (Figure 2C). Thus, influenza-specific CD8 T cells are primed in the absence of conventional lymphoid organs, albeit with a slight delay.

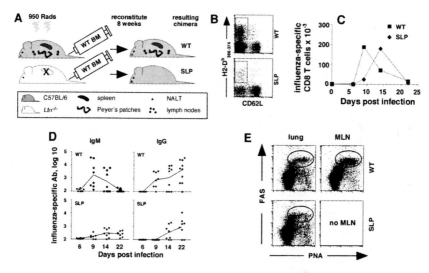

Figure 2. Respiratory immune responses are intact in the absence of conventional lymphoid organs. (A) C57BL/6 mice or splenectomized LTα$^{-/-}$ recipient mice were lethally irradiated and reconstituted with bone marrow from C57BL/6 donors. Reconstituted C57BL/6 mice are referred to as WT mice. Splenectomized and reconstituted LTα$^{-/-}$ mice lack Spleen, Lymph nodes and Peyer's patches and are referred to as SLP mice. (B) Chimeric mice were infected with influenza and nucleoprotein (NP)-specific CD8 cells were identified by tetramer binding. (C) The combined number of CD8$^+$CD62L^{lo}H–2D^{b}PA$_{224-233}$ and CD8$^+$CD62L^{lo}H–2D^{b}NP$_{366-374}$ tetramer-binding T cells in the lungs was determined by flow cytometry. (D) The serum titers of influenza-specific IgM and IgG were determined by influenza-specific ELISA. (E) CD19$^+$FAS$^+$PNA$^+$ germinal center B cells were identified in the lungs and draining lymph nodes (MLNs). The plots shown were gated on CD19$^+$ cells and the FAS^{hi}PNA^{hi} germinal center B cells are circled.

We also determined whether humoral immune responses were generated in SLP mice[29]. As shown in Figure 2D, influenza-specific IgM and IgG are produced in SLP mice, although the titers of IgM are reduced and the appearance of IgG is delayed. Since isotype switching in B cells usually occurs in germinal centers, which facilitate the proliferation and selection of B cells that produce high-affinity neutralizing antibody[30], we tested whether B cells with a germinal center phenotype could be found in SLP mice. As shown in Figure 2E, germinal center B cells are found in the lungs of both WT and SLP mice on day 14 after infection. This was somewhat surprising, as germinal centers are normally found in highly organized lymphoid tissues, such as lymph nodes or spleen. These data demonstrate that the formation of germinal center B cells and the production of influenza-specific isotype-switched antibody are not dependent on the presence of peripheral lymphoid tissues, and suggest that the lung itself is competent to initiate and maintain influenza-specific B cell responses.

2.3. Structure and Function of Bronchus-Associated Lymphoid Tissue (BALT)

The presence of B cells with a germinal center phenotype in the lungs of WT and SLP mice suggests that organized lymphoid tissues can be formed in the lung. Since the lungs of normal uninfected mice typically do not contain organized lymphoid tissue, we wanted to know whether influenza infection induces the formation of organized lymphoid structures. To test this we infected mice with influenza and analyzed thick sections of lungs for the presence of organized lymphoid tissues on day 7 post-infection. As shown in Figure 3A–B, organized lymphoid tissues are found in the lungs of mice previously infected with influenza. The majority of the cells in these areas are B cells, which are organized into follicles and are surrounded by T cells (Figure 3A–B). In areas where several B cell follicles are clustered, the interfollicular regions develop T cell zones that contain both CD4 and CD8 T cells as well as CD11c expressing dendritic cells[31]. Our published studies also show that the B cell follicles in the lung are centered on CD21-expressing follicular dendritic cells and that some B cell follicles even have well-defined germinal centers[29]. Proliferating CD8 T cells are found in the interfollicular T cell areas, while proliferating CD4 T cells are observed in the interfollicular area and in the germinal center in close proximity to rapidly proliferating B cells[29]. These data demonstrate that infection triggers the development of organized lymphoid tissues in the lung that have all of the characteristics of secondary lymphoid organs. We have termed these structures "inducible Bronchus Associated Lymphoid Tissue" or iBALT.

Although it is clear that organized lymphoid tissues can be formed in the lung and that these tissues are capable of supporting immune responses, it is not clear whether these tissues represent true secondary lymphoid organs or whether they are tertiary lymphoid tissues that are formed only in response to inflammation or infection. Bronchus Associated Lymphoid Tissue (BALT) was originally described as a mucosal lymphoid tissue found along the upper bronchi of the respiratory tract in pigs and rabbits[32,33]. In these studies, BALT appears very similar to Peyer's patches, with prominent lymphoid follicles underneath a specialized dome epithelium[33,34]. However, unlike classically defined BALT, which

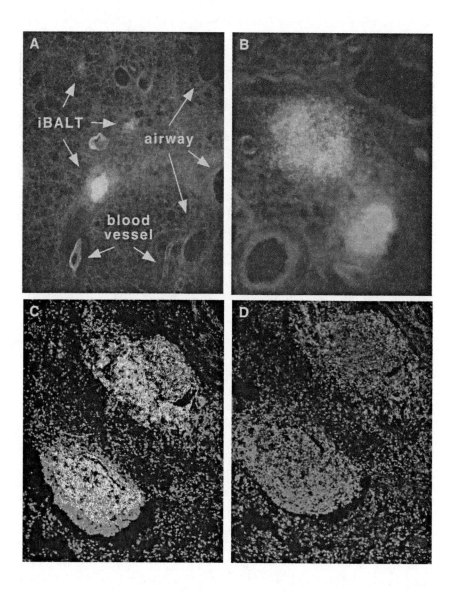

Figure 3 (see color insert). The structure of iBALT in murine and human lungs. (A–B) C57BL/6 mice were infected with influenza and 100-μM sections were prepared from lungs 3 weeks after infection and stained with antibodies to B220 (green) and CD3 (red). The brightness of the green channel was increased so that the autofluorescence of the airways and blood vessels would be visible. (C–D) Sections of a lung biopsy from a patient with follicular bronchiolitis associated with Rheumatoid Arthritis were stained with antibodies to CD20 (green) and PCNA (red) to identify proliferating B cells (C) and with antibodies to CD21 (red) to identify follicular dendritic cell networks (D). Sections C and D are counterstained with DAPI (blue).

develops during embryogenesis and is generated and maintained in the absence
of antigen[35], iBALT in both mouse and humans appears to be formed only after
infection or inflammation[36-38] and is found in numerous places along the upper
and lower bronchi and even in the interstitial areas of the lung. In fact, the spo-
radic appearance of BALT-like areas in humans and mice has led some investigators to doubt whether BALT is an important secondary lymphoid tissue in
these species[36]. However, transgenic mice that express various cytokines in the
lung develop iBALT[39,40], as do mice and humans prone to autoimmune diseases[41,42] or humans with recurrent or chronic respiratory infections[43,44]. For example, B cell follicles are easily observed in the lungs of a patient with follicular
bronchiolitis associated with Rheumatoid Arthritis (Figure 3D–E). These B cell
follicles support B cell proliferation and are centered on CD21-expressing follicular dendritic cells just as they are in murine lungs[29]. Thus, iBALT has a similar structure in both murine and human lungs, and its formation is probably triggered by similar mechanisms.

Although the cellular and molecular mechanisms that lead to iBALT formation are not known, similar ectopic lymphoid tissues often form at sites of
chronic inflammation, such as in rheumatoid joints[45-47]. Some investigators have
termed this process "lymphoid neogenesis" and suggest that chronic inflammation is responsible for the appearance of lymphoid tissues in a variety of sites[46,47].
Thus, iBALT is probably another example of an ectopic lymphoid tissue that
develops upon local inflammation triggered by infection or autoimmunity. As in
the development of conventional lymphoid organs, lymphotoxin and the homeostatic chemokines appear to play a role in the development of ectopic lymphoid
tissues[47]. For example, the transgenic expression of CXCL13, CCL21, or LTα in
the pancreas results in the development of lymphoid tissue in this organ[48-50],
complete with B and T cell areas, follicular dendritic cells (FDCs), and high
endothelial venules (HEVs). Since homeostatic chemokines are clearly important for both the development and the organization of ectopic lymphoid tissues,
we tested whether the expression of these chemokines is induced upon influenza
infection of the lung. As expected, CXCL13 mRNA is not expressed in normal
uninfected lungs[29]. However, it is rapidly induced upon infection with influenza
virus and is maintained after infection is resolved. We also demonstrated that the
expression of CXCL13 and CCL21 proteins co-localize with the B and T cell
areas observed in the lung[29]. The majority of CXCL13 protein expression local-
izes to the follicular area that contains B cells and follicular dendritic cells. In
contrast, CCL21 expression is observed on vascular endothelium as well as on
reticular cells around the edge of the follicles and in the interfollicular areas[29].
These areas are similar to where CD11c⁺ dendritic cells and T cells are found[31].
Finally, we observed that PNAd, the peripheral lymph node addressin, is expressed on vascular endothelial cells in the areas of iBALT surrounding the B
cell follicles[29]. These data demonstrate that the lymphoid areas of the lung that
form after infection are not simply accumulations of effector cells that are
primed in conventional lymphoid organs. Instead, these newly formed lymphoid

tissues express chemokines and homing receptors necessary to recruit naive lymphocytes and to organize them into structures that support lymphocyte priming and differentiation.

2.4. Does iBALT Confer Antiinflammatory Properties on Local Immune Responses?

The data above showed that B and T cell immune responses to influenza can be primed in iBALT in mice that lacked conventional lymphoid organs[29]. However, the delayed B and T cell responses observed in these mice also suggest that immunity in these mice may not be as effective in the absence of normal secondary lymphoid organs. In fact, viral clearance is effective, but somewhat delayed in mice lacking lymphoid tissues[29]. Interestingly, the delayed viral clearance did not result in increased morbidity and mortality, and SLP mice actually survived higher doses of influenza significantly better than did WT mice[29]. Furthermore, when infected with lower doses of virus that both WT and SLP mice could survive (240 and 50 EIU), the SLP mice exhibited substantially less morbidity throughout the experiment. Thus, the lack of peripheral lymphoid organs does not result in increased susceptibility to influenza but, instead, significantly increases the ability of these mice to survive infection with normally lethal doses of virus. These data are very surprising and somewhat difficult to explain. If the SLP mice make delayed immune responses and are less efficient at clearing virus, why do they exhibit less morbidity and mortality? One possibility is that, because the overall magnitude of the immune response is smaller in SLP mice, there is less immunopathology. Another possibility is that lymphocytes primed exclusively in iBALT produce antiinflammatory cytokines. These possibilities are currently being investigated.

3. CONCLUSIONS AND FUTURE DIRECTIONS

The data presented here suggest that the current paradigm, in which primary adaptive immune responses are generated exclusively in conventional lymphoid organs, such as spleen, lymph node and Peyer's patches, should be modified to include a role for locally induced lymphoid tissues, such as iBALT. In this new model, infection and inflammation activate local APCs, which traffic to pre-existing lymphoid organs and initiate primary immune responses. At the same time, infection and inflammation trigger the formation of local lymphoid tissues, such as iBALT, at the site of inflammation (Figure 4). These local lymphoid tissues are fully capable of initiating primary immune responses and of expanding effector cells that were primed in conventional lymphoid organs. The generation of primary immune responses in these local tissues may be slightly delayed relative to that in conventional lymphoid organs, because it takes time for the local tissues to develop the appropriate architecture that will recruit and prime naive lymphocytes. However, once local lymphoid tissues are formed, we predict that they will initiate primary immune responses to new antigens even faster than conventional lymphoid organs due to their close proximity to antigen. In the case of immune responses to infectious agents, this faster response will likely be beneficial. However, in the case of immune responses to local autoantigens, faster responses may be harmful and may exacerbate local pathology.

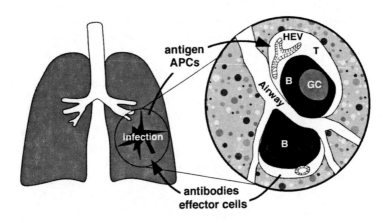

Figure 4. Local infection in the lung triggers the development of iBALT. iBALT has the structure of an ectopic lymphoid tissue, with B cell follicles, germinal centers, HEVs, and T cell areas. In concert with conventional lymphoid organs, iBALT acts as a secondary lymphoid tissue that facilitates local immune responses.

Despite the demonstration that local lymphoid tissues like iBALT are functional, numerous questions remain about how they actually work. For example, do lymphocytes primed in iBALT have different effector functions than those primed in the lymph nodes that drain the lung? Can APCs traffic from the airways directly to areas of iBALT? Do lymphocytes primed in iBALT traffic directly into the airways, or do they have to exit iBALT via efferent lymphatics and re-enter the lung from the blood? What are the mechanisms that trigger iBALT formation? How long does iBALT persist in the lung after the initial inflammation is resolved? Finally, does the presence of iBALT in human lungs provide an overall benefit or detriment to the individual?

4. ACKNOWLEDGMENTS

The authors would like to thank Dr. Frances Lund for her critical review of this manuscript and for helpful discussions during the course of this work. We gratefully acknowledge financial support for this work from NIH grants HL-69409, HL63925, the Sandler Program for Asthma Research, and the Trudeau Institute.

5. REFERENCES

1. E.C. Butcher and L.J. Picker. Lymphocyte homing and homeostasis. *Science* **272**:60–62 (1996).
2. J.G. Cyster. Chemokines and the homing of dendritic cells to the T cell areas of lymphoid organs. *J Exp Med* **189**:447–450 (1999).
3. K.L. Legge and T.J. Braciale. Accelerated migration of respiratory dendritic cells to the regional lymph nodes is limited to the early phase of pulmonary infection. *Immunity* **18**:265–277 (2003).
4. C.C. Goodnow. Chance encounters and organized rondezvous. *Immunol Rev* **156**:5–10 (1997).
5. M. Balazs, F. Martin, T. Zhou and J. Kearney. Blood dendritic cells interact with splenic marginal zone B cells to initiate T-independent immune responses. *Immunity* **17**:341–352 (2002).
6. M.R. Neutra, A. Frey and J.P. Kraehnenbuhl. Epithelial M cells: gateways of mucosal infection and immunization. *Cell* **86**:345–348 (1996).
7. P.D. Rennert, J.L. Browning, R.E. Mebius, F. Mackay and P.S. Hochman. Surface lymphotoxin a/b complex is required for the development of peripheral lymphoid organs. *J Exp Med* **184**:1999–2006 (1996).
8. P. de Togni, J. Goellner, N.H. Ruddle, P.R. Streeter, A. Fick, S. Mariathasan, S.C. Smith, R. Carlson, L.P. Shornick and J. Strauss–Schoenberger. Abnormal development of peripheral lymphoid organs in mice deficient in lymphotoxin. *Science* **264**:703–707 (1994).
9. P.A. Koni, R. Sacca, P. Lawton, J.L. Browning, N.H. Ruddle and R.A. Flavell. Distinct roles in lymphoid organogenisis for lymphotoxins a and b revealed in lymphotoxin b deficient mice. *Immunity* **6**:491–500 (1997).
10. A. Futterer, K. Mink, A. Luz, M.H. Kosco-Vilbois and K. Pfeffer. The lymphotoxin b receptor controls organogenisis and affinity maturation in peripheral lymphoid tissues. *Immunity* **9**:59–70 (1998).
11. S. Miyawaki, Y. Nakamura, H. Suzuka, M. Koba, R. Yasumizu, S. Ikehara and Y. Shibata. A new mutation, aly, that induces a generalized lack of lymph nodes accompanied by immunodeficiency in mice. *Eur J Immunol* **24**:429–434 (1994).
12. K. Honda, H. Nakano, H. Yoshida, S. Nishikawa, P. Rennert, K. Ikuta, M. Tamechika, K. Yamaguchi, T. Fukumoto, T. Chiba and S.I. Nishikawa. Molecular basis for hematopoietic/mesenchymal interaction during initiation of Peyer's patch organogenesis. *J Exp Med* **193**:621–630 (2001).
13. S.A. Luther, K.M. Ansel and J.G. Cyster. Overlapping roles of CXCL13, interleukin 7 receptor alpha, and CCR7 ligands in lymph node development. *J Exp Med* **197**:1191–1198 (2003).
14. K.M. Ansel, V.N. Ngo, P.L. Hayman, S.A. Luther, R. Forster, J.D. Sedgwick, J.L. Browning, M. Lipp and J.G. Cyster. A chemokine-driven positive feedback loop organizes lymphoid follicles. *Nature* **406**:309–314 (2000).

15. M.D. Gunn, K. Tangemann, C. Tam, J.D. Cyster, S.D. Rosen and L.T. Williams. A chemokine expressed in lymphoid high endothelial venules promotes the adhesion and chemotaxis of naive T lymphocytes. *Proc Natl Acad Sci* **95**:258–263 (1998).

16. M.D. Gunn, S. Kyuwa, C. Tam, T. Kakiuchi, A. Matsuzawa, L.T. Williams and H. Nakano. Mice lacking expression of secondary lymphoid organ chemokine have defects in lymphocyte homing and dendritic cell localization. *J Exp Med* **189**:451–460 (1999).

17. V.N. Ngo, H. Korner, M.D. Gunn, K.N. Schmidt, D.S. Riminton, M.D. Cooper, J.L. Browning, J.D. Sedgewick and J.G. Cyster. Lymphotoxin a/b and tumor necrosis factor are required for stromal cell expression of homing chemokines in B and T cell areas of the spleen. *J Exp Med* **189**:403–412 (1999).

18. S.L. Constant, J.L. Brogdon, D.A. Piggott, C.A. Herrick, I. Visintin, N.H. Ruddle and K. Bottomly. Resident lung antigen-presenting cells have the capacity to promote Th2 T cell differentiation in situ. *J Clin Invest* **110**:1441–1448 (2002).

19. F.E. Lund, S. Partida-Sanchez, B.O. Lee, K.L. Kusser, L. Hartson, R.J. Hogan, D.L. Woodland and T.D. Randall. Lymphotoxin-alpha-deficient mice make delayed, but effective, t and b cell responses to influenza. *J Immunol* **169**:5236–5243 (2002).

20. F. Amiot, P. Vuong, M. Desfontaines, C.D. Pater and F.M. Liance. Secondary alveolar echinococcosis in lymphotoxin-a and tumour necrosis factor-a deficient mice: exacerbation of *Echinococcus multilocularis* larval growth is associated with cellular changes in the periparasitic granuloma. *Parisite Immunol* **21**:475–483 (1999).

21. T.A. Banks, B.T. Rouse, M.K. Kerley, P.J. Blair, V.L. Godfrey, N.A. Kuklin, D.M. Bouley, J. Thomas, S. Kanangat and M.L. Mucenski. Lymphotoxin a deficient mice: effects on secondary lymphoid organ development and humoral immune responsiveness. *J Immunol* **155**:1685–1693 (1995).

22. D.P. Berger, D. Naniche, M.T. Crowley, P.A. Koni, R.A. Flavell and M.B.A. Oldstone. Lymphotoxin-b-deficient mice show defective antiviral immunity. *Virology* **260**:136–147 (1999).

23. U. Kumaraguru, I.A. Davis, S. Deshpande, S.S. Tevethis and B.T. Rouse. Lymphotoxin a$^{-/-}$ mice develop functionally impaired CD8+ t cell responses and fail to contain virus infection of the central nervous system. *J Immunol* **166**:1066–1074 (2001).

24. B.J. Lee, S. Santee, S. Von Gesjen, C.F. Ware and S.R. Sarawar. Lymphotoxin-a-deficient mice can clear a productive infection with murine gammaherpesvirus 68 but fail to develop spenomegaly or lymphocytosis. *J Virol* **74**:2786–2792 (2000).

25. A. Harmsen, K. Kusser, L. Hartson, M. Tighe, M.J. Sunshine, J.D. Sedgwick, Y. Choi, D.R. Littman and T.D. Randall. Organogenesis of Nasal Associated Lymphoid Tissue (NALT) occurs independently of lymphotoxin-α (LTα) and retinoic acid receptor-related orphan receptor-γ, but the organization of NALT is LTα-dependent. *J Immunol* **168**:986–990 (2002).

26. J. Rangel-Moreno, J. Moyron-Quiroz, K. Kusser, L. Hartson, H. Nakano and T.D. Randall. Role of CXC chemokine ligand 13, CC chemokine ligand (CCL) 19, and CCL21 in the organization and function of nasal-associated lymphoid tissue. *J Immunol* **175**:4904–4913 (2005).

27. M. Matsumoto, S. Mariathasan, M.H. Nahm, F. Baranyay, J.J. Peschon and D.D. Chaplin. Role of lymphotoxin and the type 1 TNF receptor in the formation of germinal centers. *Science* **271**:1289–1291 (1996).

28. M. Matsumoto, S.F. Lo, C.J.L. Carruthers, J. Min, S. Mariathasan, G. Huang, D.R. Plas, S.M. Martin, R.S. Geha, M.H. Nahm and D.D. Chaplin. Affinity maturation without germinal centers in lymphotoxin a deficient mice. *Nature* **382**:462–466 (1996).

29. J.E. Moyron-Quiroz, J. Rangel-Moreno, K. Kusser, L. Hartson, F. Sprague, S. Goodrich, D.L. Woodland, F.E. Lund and T.D. Randall. Role of inducible bronchus associated lymphoid tissue (iBALT) in respiratory immunity. *Nat Med* **10**:927–934 (2004).

30. G. Kelsoe. The germinal center: a crucible for lymphocyte selection. *Semin Immunol* **8**:179–184 (1996).

31. D.L. Woodland and T.D. Randall. Anatomical features of anti-viral immunity in the respiratory tract. *Semin Immunol* **16**:163–170 (2004).

32. J. Bienenstock, N. Johnston and D.Y. Perey. Bronchial lymphoid tissue, I: morphologic characteristics. *Lab Invest* **28**:686–692 (1973).

33. J. Bienenstock and N. Johnston. A morphologic study of rabbit bronchial lymphoid aggregates and lymphoepithelium. *Lab Invest* **35**:343–348 (1976).

34. T. Sminia, G.J. van der Brugge-Gamelkoorn and S.H. Jeurissen. Structure and function of bronchus-associated lymphoid tissue (BALT). *Crit Rev Immunol* **9**:119–150 (1989).

35. S. Delventhal, A. Hensel, K. Petzoldt and R. Pabst. Effects of microbial stimulation on the number, size and activity of bronchus-associated lymphoid tissue (BALT) structures in the pig. *Int J Exp Path* **73**:351–357 (1992).

36. R. Pabst and I. Gehrke. Is the bronchus-associated lymphoid tissue (BALT) an integral structure of the lung in normal mammals, including humans? *Am J Respir Cell Mol Biol* **3**:131–135 (1990).

37. I. Richmond, G.E. Pritchard, T. Ashcroft, A. Avery, P.A. Corris and E.H. Walters. Bronchus associated lymphoid tissue (BALT) in human lung. its distribution in smokers and non-smokers:*Thorax* **48**, 1130–1134 (1993).

38. T. Tshering and R. Pabst. Bronchus associated lymphoid tissue (BALT) is not present in normal adult lung but in different diseases. *Pathobiology* **68**:1–8 (2000).

39. J. J. Lee, M.P. McGarry, S.C. Farmer, K.L. Denzler, K.A. Larson, P.E. Carrigan, I.E. Brenneise, M.A. Horton, A. Haczku, E.W. Gelfan, G.D. Leikauf and N.A. Lee. Interleukin-5 expression in the lung epithelium of transgenic mice leads to pulmonary changes pathognomonic of asthma. *J Exp Med* **185**:2143–2156 (1997).

40. S. Goya, H. Matsuoka, M. Mori, H. Morishita, H. Kida, Y. Kobashi, T. Kato, Y. Taguchi, T. Osaki, I. Tachibana, N. Nishimoto, K. Yoshizaki, I. Kawase, and S. Hayashi. Sustained interleukin-6 signalling leads to the development of lymphoid organ-like structures in the lung. *J Pathol* **200**:82–87 (2003).

41. B. Xu, N. Wagner, L.N. Pham, V. Magno, Z. Shan, E.C. Butcher and S.A. Michie. Lymphocyte homing to bronchus-associated lymphoid tissue (BALT) is mediated by L-selectin/PNAd, alpha4beta1 integrin/VCAM-1, and LFA-1 adhesion pathways. *J Exp Med* **197**:1255–1267 (2003).

42. R.K. Chin, J.C. Lo, O. Kim, S.E. Blink, P.A. Christiansen, P. Peterson, Y. Wang, C. Ware and Y.X. Fu. Lymphotoxin pathway directs thymic Aire expression. *Nat Immunol* **4**:1121–1127 (2003).

43. H.D. Chen, A.E. Fraire, I. Joris, M.A. Brehm, R.M. Welsh and L.K. Selin. Memory CD8+ T cells in heterologous antiviral immunity and immunopathology in the lung. *Nat Immunol* **2**:1067–1076 (2001).

44. J.H. Vernooy, M.A. Dentener, R.J. van Suylen, W.A. Buurman and E.F. Wouters. Long-term intratracheal lipopolysaccharide exposure in mice results in chronic lung inflammation and persistent pathology. *Am J Respir Cell Mol Biol* **26**:152–159 (2002).

45. N.H. Ruddle. Lymphoid neo-organogenesis: lymphotoxin's role in inflammation and development. *Immunol Res* **19**:119–125 (1999).

46. A. Kratz, A. Campos-Neto, M.S. Hanson and N.H. Ruddle. Chronic inflammation caused by lymphotoxin is lymphoid neogenesis. *J Exp Med* **183**:1461–1472 (1996).

47. P. Hjelmstrom, J. Fjell, T. Nakagawa, R. Sacca, C. Cuff and N. Ruddle. Lymphoid tissue homing chemokines are expressed in chronic inflammation. *Am J Pathol* **156**:1133–1138 (2000).

48. S.A. Luther, T. Lopez, W. Bai, D. Hanahan and J.G. Cyster. BLC expression in pancreatic islets causes b cell recruitment and lymphotoxin-dependent lymphoid neogenesis. *Immunity* **12**:471–481 (2000).

49. S.A. Luther, A. Bidgol, D.C. Hargreaves, A. Schmidt, Y. Xu, J. Paniyadi, M. Matloubian and J.G. Cyster. Differing activities of homeostatic chemokines CCL19, CCL21, and CXCL12 in lymphocyte and dendritic cell recruitment and lymphoid neogenesis. *J Immunol* **169**:424–433 (2002).

50. R. Sacca, C.A. Cuff, W. Lesslauer and N.H. Ruddle. Differential activities of secreted lymphotoxin-alpha3 and membrane lymphotoxin-alpha1beta2 in lymphotoxin-induced inflammation: critical role of TNF receptor 1 signaling. *J Immunol* **160**:485–491 (1998).

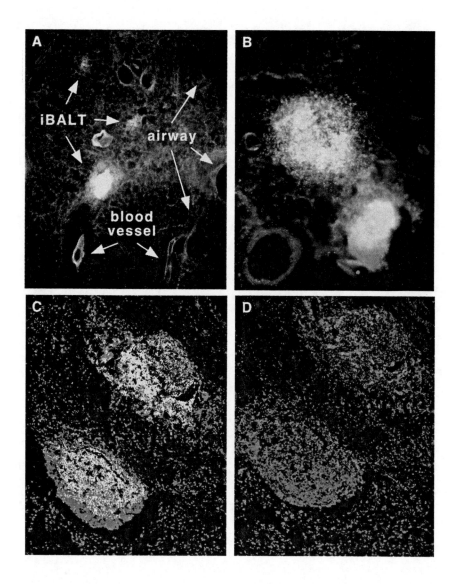

Figure 4.3. The structure of iBALT in murine and human lungs. (A–B) C57BL/6 mice were infected with influenza and 100-μM sections were prepared from lungs 3 weeks after infection and stained with antibodies to B220 (green) and CD3 (red). The brightness of the green channel was increased so that the autofluorescence of the airways and blood vessels would be visible. (C–D) Sections of a lung biopsy from a patient with follicular bronchiolitis associated with Rheumatoid Arthritis were stained with antibodies to CD20 (green) and PCNA (red) to identify proliferating B cells (C) and with antibodies to CD21 (red) to identify follicular dendritic cell networks (D). Sections C and D are counterstained with DAPI (blue).

5

THE YIN AND YANG OF ADAPTIVE IMMUNITY IN ALLOGENEIC HEMATOPOIETIC CELL TRANSPLANTATION: DONOR ANTIGEN-PRESENTING CELLS CAN EITHER AUGMENT OR INHIBIT DONOR T CELL ALLOREACTIVITY

Jian-Ming Li and Edmund K. Waller

The immunoregulatory activity of different donor bone marrow (BM) cell subsets has not yet been fully addressed in allogeneic transplantation. We studied whether manipulation of donor antigen-presenting cells (APC) can affect post-transplant immunity using a mouse model of allogeneic bone marrow transplantation (BMT). CD11b is a marker present on mature monocytes, granulocytes, and a subset of dendritic cells (DC). In order to manipulate the content of APC, we enriched or depleted $CD11b^+$ cells from BM grafts using immuno-magnetic cell sorting. The effect of CD11b depletion on graft-versus-host disease (GvHD) and graft-versus-leukemia (GvL) was studied in a MHC fully mismatched model of allogeneic BMT using C57BL/6 $\rightarrow$ B10.BR transplants and LBRM cells, a B10.BR T cell leukemia cell line. Transplantation with CD11b partially or fully depleted BM and low-dose donor splenocytes conferred 40% long-term leukemia- free survival with minimal GvHD when supralethal doses of LBRM were administered before transplant, or 75 days after BMT. Higher levels of serum gamma interferon and expansion of spleen-derived $CD4^+$ memory T cells were seen among recipients of CD11b-depleted BM compared to recipients of unmanipulated BM. Expansion of donor-spleen-derived T cells was inversely proportional to the content of $CD11b^+$ cells in the BM graft. Thus, manipulating the content of APC subsets in donor BM by enriching or removing $CD11b^+$ cells had a direct effect on post-transplant immunity and the balance between donor T cell activation (Yang) and donor T cell tolerance/anergy (Yin). Thus an optimal balance between the opposing states of donor T cell activation and anergy can be achieved in allogeneic BMT through the immunoregulatory activities of donor DC.

Jian-Ming Li, Edmund K. Waller. Emory University, Atlanta, GA 30322, USA.

1. INTRODUCTION

Allogeneic hematopoietic progenitor cell transplant (HPCT) is designed to re-place host hematopoiesis with donor-derived hematopoietic stem cells and their progeny. The network of interactions that is responsible for innate and adaptive immune responses ultimately results in reconstitution of immunity following transplant. At the moment donor T cells are infused to a transplant recipient that has been conditioned with myeloablative doses of radiation, recipient type anti-gen-presenting cells (APC) are still present and can induce allo-activation of donor T cells that leads to graft-versus-host reactions[1]. Donor APC that are in-fused with the graft also have a role in donor T cell activation, and can influence graft-versus-host reactions[2]. In contrast to the ability of donor and host APC to induce donor T cell alloreactivity, the role of APC that inhibit or suppress donor T cell activation is less defined. We have previously described a clinical associa-tion between the content of donor dendritic cells (DC) and post-transplant graft-versus-host disease (GvHD) and leukemic relapse[3]. Recipients of larger numbers of plasmacytoid DC precursors had a higher incidence of post-transplant relapse and less chronic GvHD, suggesting that indirect presentation of alloantigen by donor APC (or their progeny) to donor T cells was responsible for suppressing innate and/or adaptive immune responses that lead to alloreactive donor T cells[3]. A recent report by Reddy et al. indicated that the content of donor-derived DC precursors in the blood at the time of hematopoietic engraftment was strongly associated with incidence of relapse, GvHD, and overall post-transplant sur-vival[4]. Thus donor DC precursors in the graft and donor-derived DC in the early post-transplant period appear to be important in regulating GvHD and GvL ef-fects in clinical outcomes following BMT in patients with hematological malig-nancies[3,4]. These clinical data suggest a relationship between donor DC subtypes and their capacity to regulate activation of alloreactive T cells in response to indirect presentation of alloantigen.

In-vitro experiments using purified populations of human DC have demon-strated that phenotypically defined DC subsets can direct naive T cells toward either Th1 or Th2 differentiation[5,6]. So-called "myeloid" human DC direct naive T cells toward Th1 polarization[7], while so-called "lymphoid" or "plasmacytoid" human DC direct naive T cells toward Th2 polarization[8,9]. However, the precise role of different DC subsets in post-transplant immunity remains undefined, and the relationship between phenotypically defined DC subsets and their capacity to direct the immune polarization of naive T cells in vivo is unclear[10,11].

In contrast to human DC, all murine DC appear to express CD11c (the marker for human DC1). Murine CD11c$^+$ DC and DC precursors can be divided into two groups: one that lacks CD11b expression (CD8α^+ DEC-205$^+$, CD4$^-$ CD8α^{low} DEC-205$^+$, and CD4$^-$CD8α^- DEC-205$^+$ subsets), and a second group that expresses CD11b (CD11b$^+$CD4$^-$CD8α^- DEC-205$^-$, and CD11b$^+$ CD4$^+$ CD8α^- DEC-205$^-$)[12]. The murine DC that lack CD11b expression but express B220$^+$ (CD4$^-$CD8α^-DEC-205$^+$)[13] have phenotypic homology to human plasmacytoid DC2 and, like human pDC2, synthesize large amounts of interferon alpha in response to viral infection or binding of CpG sequences to TLR 9[14]. Pelayo et al. have defined two subsets of murine plasmacytoid DC progenitors. The first, lacking Rag promoter activity, induces cognate T cells to make large amounts of IL12 p70 in response to activation signals of SAC, CPG, and LPS plus CD40L, and triggers Th1 responses, while the second, utilizing the Rag promoter, induces T cells toward Th2 immune responses[15]. Both CD11b$^-$/CD8α^+ and CD11b$^+$/CD8α^- murine DC efficiently stimulate T cell responses to protein-antigen or alloantigen in vitro and in vivo[16,17]. But other reports have shown that murine CD11b$^+$/CD8α^- DC polarize T cells toward Th2 or mixed Th1/Th2 responses[5,6]. The CD11b$^+$ myeloid DC have shown suppressive activity mediated through nitric oxide synthesis and dependent upon IFN-γ[18].

We have recently developed a murine model of bone marrow transplantation to explore the relationship between different donor DC subsets and transplant outcomes that is instructive in defining the role of donor APC in regulating post-transplant adaptive cellular immunity[19]. In order to explore the role of DC (and DC precursor) subsets in allogeneic BMT, we manipulated the content of CD11b$^+$ cells (predominantly containing immature CD11b$^+$DC) in BM grafts in a murine allogeneic BMT model.[19] We hypothesized that the relative balance between donor T cell activation (Yang) and donor T cell tolerance (Yin) in post-transplant immunity may be regulated, in part, by the relative numbers of DC subsets in the BM graft (Figure 1), and used CD11b as a convenient antigenic marker to enrich or deplete DC subsets using immuno-magnetic cell sorting (MACs). We measured post-transplant expansion of congenically marked allogeneic donor T cells co-transplanted with BM cells and the GvL activity of donor T cells to test the hypothesis that altering the content of donor APC in the BM graft would enhance donor T cell activation (Yang) during post-transplant immune reconstitution. The data confirm the role of donor APC in regulating the Yin and Yang of adaptive cellular immune responses following allogeneic BMT.

2.   MATERIALS AND METHODS

2.1.   Mice

B10.BR (H-2k^k) and B6.SJL (CD90.2) CD45.1 congenic strains of C57.BL6 (H-2k^b) (CD45.2, CD90.2) mice aged 8 to 10 weeks were purchased from Jackson Laboratories (Bar Harbor, ME). Congenic strains expressing CD90.1 and CD45.1 on a C57.BL6 (H-2k^b) background, and CD90.1, CD45.2 on a B10.BR (H-2k^k) background were bred at the Emory University Animal Care Facility.

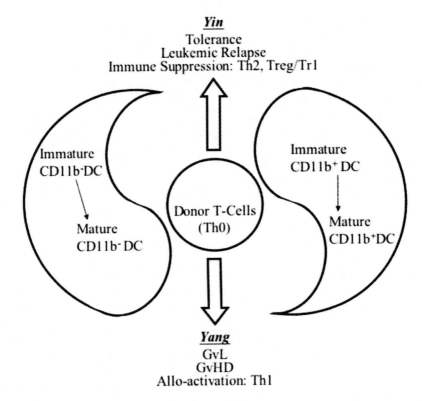

Figure 1. Diagram of Yin and Yang. Phenotype and function of DC and DC precursor subsets in freshly isolated BM cells. Phenotypically defined, CD11bˉDC and precursors are listed on the left, CD11ˉDC and precursors are listed on the right. Functionally, mature CD11bˉDC and CD11bˉDC stimulate Th0 development into Th1, as shown on the lower section of the diagram, and immature CD11bˉDC and CD11bˉDC and precursors polarize Th0 development into Th2 or Treg or Tr1.

2.2. LBRM Tumor Cell Line

The mouse LBRM–33 5A4 T lymphoma cell line[20] was purchased from American Type Culture Collection (ATCC) and assayed to be free of LCMV, MHV, MMV, and MPV, etc. by PCR-based screening tests performed by University of Missouri Research Animal Diagnostic Laboratory. LBRM were passaged according to ATCC instructions.

2.3. Donor Cell Preparations

Femora and tibia were isolated from donor mice. BM cells were harvested by flushing the cells out of the bone shaft with sterile RPMI1640 containing 1%

fetal calf serum (RPMI/FCS). Splenocytes were harvested by perfusing the spleen with sterile RPMI/FCS and the cell suspension filtered to remove cell clumps and debris.

2.4. BM CD11b Depletion and Splenic T Cell Purification

BM cells were stained with biotinylated anti-CD11b antibody, washed once in RPMI/FCS, and then resuspended with Streptavidin Microbeads (Miltenyi Biotech Gmbh, Bergisch Gladbach, Germany). The CD11b⁻ cell fraction in BM was separated from the CD11b⁺ cell fraction using the Vario MACs magnetic separation column C. The average efficiency of depletion was 99.9%. The average purity of CD11b-enriched BM cells was $87 \pm 3\%$. The partially CD11b-depleted BM graft consisted of a 50:50 mixture of unmanipulated BM and fully CD11b-depleted BM. For isolation of T cells, splenocytes were stained with biotinylated anti-CD11b, B220, DX5, and TER119 antibodies and Streptavidin Microbeads. Non-bound cells, enriched for CD3⁺ T cells, were isolated from splenocytes by an LS column in the Vario MACs separation device (Miltenyi Biotech, Gmbh). Average purity of CD3⁺ T cells was $93 \pm 2\%$ in the non-bound fraction.

2.5. Recipient Mice Conditioning

We used a standard myeloablative conditioning regimen that we have previously described[21,22]. Briefly, two days before BMT, recipient mice were exposed to two equal 5.5-Gy fractions 3 hours apart at a dose rate of 1.24 Gy/min[21,23]. Lethally irradiated mice were maintained on oral aqueous antibiotics (1.1 mg/mL neomycin sulfate and 1000 U/mL polymyxin sulfate) following irradiation and for 4 weeks after BMT.

2.6. BMT and Leukemia Challenge

BMT and leukemia infusions were performed by injecting 0.2 mL cell suspensions into the tail vein of recipient mice. Recipient mice were challenged with ten times the lethal dose of LBRM tumor cells (3×10^6) on day 1 (one day prior to transplantation) or day +75 post-transplantation. BM, splenocytes, or purified T cells were injected jointly on day 0 (two days after TBI).

2.7. Analyses of DC Subsets and DC Precursors in BM and Spleen Grafts

The content of DC subsets and DC precursors in freshly isolated BM and splenocytes was assessed by staining with a cocktail of monoclonal antibodies to differentiate hematopoietic lineage-associated antigens (CD3, CD19, DX5, sIgM, and TER119) from markers of subsets of DC and DC precursors (CD4, CD8α, CD11b, CD11c, and B220)[19].

2.8. Analyses of Hematopoietic Engraftment of Transplant Recipients

0.2 mL peripheral blood was collected from the tail vein at days +30, +60, and +105 after transplant. Red blood cells were depleted by ammonium chloride lysis. Flow cytometry analysis was performed as we have previously described[19] using monoclonal antibodies to distinguish between various leukocyte lineages as well as congenic markers to distinguish CD4$^+$ and CD8$^+$ T cells derived from donor spleen (CD90.1$^+$) or donor BM (CD90.2$^+$). CD4$^+$ CD62L$^-$ CD44^{high} cells were defined as CD4 memory T cells; CD8$^+$ Ly-6c$^+$ CD44^{high} cells were defined as CD8 memory T cells.[22,24–26] Absolute numbers of CD4 or CD8 memory T cell subsets in blood were calculated by multiplying the fraction of blood leukocytes with a memory subpopulation phenotype by absolute leukocyte concentration.

2.9. Serum Gamma Interferon (IFN-γ) and Tumor Necrosis Factor-Alpha (TNF-α) Enzyme-Linked Immunosorbent Assay (ELISA)

Serum samples were obtained from mouse-tail vein blood using microtainer serum separator tubes (Becton Dickinson, Franklin Lakes, NJ). The serum was analyzed for IFN-γ and TNF-α using commercial IFN-γ and TNF-α OptEIA ELISA kits (Pharmingen, San Diego, CA), with sensitivities of 14 and 5 pg/ml, respectively, according to the manufacturer's procedures. Reaction plates were analyzed using a SpectraMax 340PC spectrophotometer (Molecular Devices, Sunnyvale, CA).

2.10. Assessments of Survival and GvHD in Transplant Recipients

All transplant recipients were evaluated for clinical signs of GvHD by weight loss, posture, activity, fur texture, and skin integrity[27] once or twice weekly. Survival was monitored daily. Moribund mice, animals that lost more than 25% body weight, and mice surviving to the end of the experiment (day 100 in the absence of added tumor cells; day 150 for tumor administered on day −1; and day 220 for tumor administered on day +75) were euthanized and submitted for necropsy (minimum of 5 mice/group). The cause of death was determined by necropsy (either GvHD or leukemia, when histological evidence of either condition was present). Slides of the liver and small intestine were coded without knowledge of treatment conditions, and scored for evidence of GvHD (grades 0–4) using previously reported criteria[21,28–30].

2.11. Statistical Analyses

Survival differences between different groups were calculated with the Kaplan-Meier log rank test in a pairwise fashion[21]. Differences of GvHD score among the groups were compared using the Mann-Whitney U-test[27]. Other data, such as levels of donor T cells among groups, were compared using one-way ANOVA.

3.   RESULTS

3.1.   MACs Depletion of CD11b$^+$ Cells in the BM Graft Does not Affect Stem Cell Content

In a model system of MHC-fully mismatched BMT, we used allogeneic grafts consisting of a mixture of 5×10^6 BM cells and 1×10^6 splenocytes from C57BL/6 donors (H-2k^b) transplanted into lethally irradiated B10.BR (H-2k^k) recipients. This dose of BM was sufficient to prevent graft rejection, and 1×10^6 splenocytes is below the threshold number that leads to lethal GvHD in this donor-recipient strain combination. We calculated the total number of DC subsets contained in the mixture of BM cells and splenocytes, in which the BM component of the graft was either unmanipulated (BM); CD11b-depleted (CD11b [–] BM); partially CD11b-depleted (CD11b [+/–] BM, representing a 50:50 combination of unmanipulated BM and CD11b-depleted BM); or CD11b enriched (CD11b [+] BM). The numbers of CD11b$^+$ DC and DC precursors were significantly lower in the CD11b [–] BM or CD11b [+/–] BM grafts, while CD11b$^-$ DC and precursors were relatively enriched in the grafts containing CD11b [–] BM or CD11b [+/–] BM (Table 1). Depletion of CD11b$^+$ cells from the BM graft did not result in a significant depletion of total DC/DC precursor content in the graft, had no significant effect on the number of phenotypically defined multipotent HPC ("stem cells"), and did not produce a significant change in the number of NK cells in the graft[19].

3.2.   Transplanting Manipulated BM Grafts in the Absence of Added Splenocytes Did not Lead to Graft Rejection or GvHD

B10.BR mice were lethally irradiated and transplanted with 5×10^6 unmanipulated BM or 5×10^6 of CD11b [+/–] BM, CD11b[–] BM, or CD11b[+] BM. All B10.BR recipients of unmanipulated or manipulated BM alone survived[19] and did not experience GvHD that was evident by either persistent weight loss or clinical signs through day +100 post-transplant (Figure 2B). The degree of donor chimerism among recipients of these different donor BM grafts was >99% at all time points studied.[19]

3.3.   CD11b -Depleted BM Grafts Combined with Low-Dose Splenocytes or Splenic T Cells Led to Slight Enhancement of Non-Lethal GvHD in Recipients of Allogeneic BMT

The GvHD potential of graded doses of donor splenocytes combined with BM grafts depleted of CD11b$^+$ cells was determined in the C57BL/6 → B10.BR

Table 1. Content of DC Subsets in CD11b-Manipulated BM
and Splenocyte Graft Constituents

Graft	Lin⁻CD11c⁺CD11b⁺ (x10³)		Lin⁻CD11c⁺CD11b⁻ (x10³)		
	B220⁻		B220⁺		B220⁻
	CD4⁻CD8⁻	CD4⁺	CD4⁻CD8⁻	CD4⁺	CD8⁺
BM	24.8 ± 2.5	3.3 ± 0.5	17.6 ± 2.5	15.6 ± 2.5	4.1 ± 2.0
CD11b[+] BM	42.1 ± 4.0*	5.1 ± 0.1*	10.5 ± 1.0*	9.6 ± 0.5	2.6 ± 0.5
CD11b[+/−] BM	12.7 ± 1.0*	1.6 ± 0.0*	29.2 ± 2.0*	27.5 ± 1.5	9.2 ± 2.0*
CD11b[−] BM	<1	<1	41.0 ± 4.0*	39.6 ± 4.0*	13.9 ± 1.5
Unmanipulated SP	5.9 ± 0.5	14.0 ± 1.3	0.9 ± 0.1	1.6 ± 0.1	3.9 ± 0.4

Data represent the cell numbers of CD11b⁻ or CD11b⁻ DC cell subsets among the overall hematopoietic (nucleated) cells of the 5 x 10⁶ BM and 1 x 10⁶ splenocyte grafts. BM grafts were unmanipulated (BM), CD11b-enriched (CD11b[+]), partially CD11b-depleted (CD11b[+/−]), or fully CD11b-depleted (CD11b[−]). Mean value and standard deviation for each group are shown, $n = 5$ separate samples of BM and spleen). Significant differences (*$p < 0.05$) of the cell populations are shown compared to the same populations in unmanipulated BM.

BMT model. Separate groups of mice were transplanted with unmanipulated BM, CD11b [+/−] BM, or CD11b [−] BM in combination with graded cell doses of donor splenocytes (0.5×10^6, 1×10^6, 5×10^6, or 10×10^6). 80% of recipients of 5×10^6 and all recipients of 10×10^6 splenocytes died within the first 60 days post-transplant with clinical (Figure 2A,E,F) and histopathologic signs of GvHD (data not shown). In contrast, the recipients of low-dose splenocytes (0.5×10^6 or 1×10^6) had over 90% survival at day +100 post-transplant among all transplant groups[19] (Figure 2A). There were no significant differences in the incidence or severity of GvHD between the recipients of unmanipulated BM and CD11b [−] BM that received the same doses, $\geq 1 \times 10^6$, of donor splenocytes (Figure 2D,E,F). Of note, among recipients of the lowest number of donor splenocytes (0.5×10^6), the recipients of CD11b [+/−] BM or CD11b [−] BM had slightly, but significantly more, signs of mild clinical GvHD compared with recipients of unmanipulated BM ($p < 0.05$; Figure 2C).

3.4.　CD11b⁺ Cell-Enriched BM Grafts Combined with Low-Dose Splenocytes or Splenic T Cells Inhibited GvHD in Recipients of Allogeneic BMT

In contrast to the lethal GvHD that developed among the recipients of $>1 \times 10^6$ donor splenocytes combined with either unmanipulated BM or CD11b [−] BM (Figure 2E, F), the recipients of CD11b [+] BM had significantly less GvHD (*p*

< 0.05; Figure 2D,E,F), and longer survival[19] (Figure 2F), indicating that large numbers of donor CD11b[+] cells were capable of inhibiting the allo-activation of donor splenocytes.

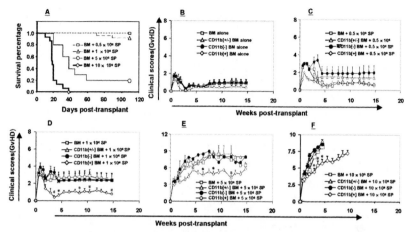

Figure 2. Survival and GvHD clinical scores of recipients transplanted with combinations of unmanipulated or manipulated BM and graded doses of donor splenocytes. Clinical signs of GvHD (weight loss, posture, activity, fur texture, and skin integrity, maximum score 10) were assessed once or twice per week up to day 100 according to the procedure listed in Materials and Methods. Survival curves were compared among the groups of mice that received unmanipulated BM grafts plus 0.5 x 10^6 splenocytes (open squares), 1 x 10^6 splenocytes (open triangles), 5 x 10^6 splenocytes (open circles), and 10 x 10^6 splenocytes (open diamonds) (panel A). Mean values for clinical GvHD score at each time point were determined in groups of mice that received unmanipulated BM grafts (open squares), CD11b[+/–] BM grafts (open triangles), CD11b[–] BM grafts (filled circles), and CD11b[+] BM grafts (open diamonds) without added splenocytes (panel B), 0.5 x 10^6 splenocytes (panel C), 1 x 10^6 splenocytes (panel D), 5 x 10^6 splenocytes (panel E), and 10 x 10^6 splenocytes (panel F). The data shown represent the overall mean values ±SD for all clinical GvHD scores at each time point for each group, and are representative of similar results seen in three separate experiments. * = $p < 0.05$ comparing the GvHD score between recipients of an unmanipulated BM graft versus CD11b [+] BM with same dose of splenocytes.

3.5.   No Combinations of Unmanipulated BM and Splenocytes Produced a GvL Effect without also Causing Lethal GvHD

We next tested the GvL and GvHD activity of various combinations of C57BL/6 BM and C57BL/6 splenocytes in B10.BR recipients that had received a supralethal dose of the T cell leukemia cell line LBRM[31] immediately after total body irradiation conditioning and before transplant. Recipients of unmanipulated BM alone or the combination of BM cells with a dose of splenocytes (1 × 10^6) that did not produce lethal GvHD when transplanted in the absence of LBRM (Figure 2A) died of progressive leukemia when 3 × 10^6 LBRM (>10 × LD_{50}) was administered one day preceding BMT (Figure 3A). Transplantation of larger numbers of donor splenocytes (5×10^6 or 10×10^6) in combination with BM resulted in rapid death of allogeneic recipients from GvHD.[19]

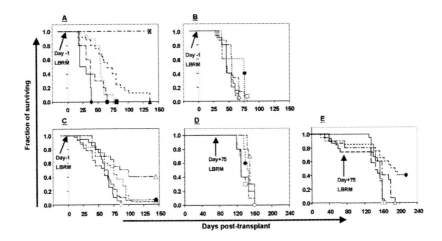

Figure 3. GvL activity of BM and spleen grafts against LBRM was enhanced by CD11b depletion of the BM graft in C57BL/6 → B10.BR transplants. B10.BR recipient mice were lethally irradiated (11 Gy) on day 2 and transplanted with various combinations of BM and splenocytes from C57BL/6 donors on day 0. The GvL effect of the allogeneic transplant was assessed by administering a lethal dose of 3×10^6 LBRM cells on day –1 (panels A, B, and C), or day +75 post-transplant (panels D and E). The content of CD11b⁺ cells in the BM graft in panels B, C, D, and E varied between unmanipulated BM (open square), CD11b[–] BM (filled circle), CD11b[+/–] BM (open triangle), or CD11b[+] BM (open diamond). Each treatment group contained 15 or 20 mice pooled from 3–4 experiments. (A) Survival of mice transplanted with LBRM on day –1 with either 5×10^6 BM alone (filled square), 5×10^6 BM with 1×10^6 splenocytes (filled triangle), 5×10^6 BM with 5×10^6 splenocytes (filled circle), and 5×10^6 BM with 10×10^6 splenocytes (filled diamond) on day 0. A control group received 1×10^6 BM and 1×10^6 splenocytes without LBRM (filled cross, dashed line). (B) Survival of mice transplanted with BM alone and LBRM on day –1. (C) Survival of mice transplanted with BM grafts and 1×10^6 splenocytes, and LBRM on day –1. (D) Survival of mice transplanted with BM alone following LBRM challenge on day +75. (E) Survival of mice transplanted with BM grafts and 1×10^6 splenocytes with LBRM administered on day +75 post-transplant. Reprinted with permission from J.M. Li and E.K. Waller. Donor antigen-presenting cells regulate T-cell expansion and antitumor activity after allogeneic bone marrow transplantation. *Biol Blood Marrow Transplant* **10**(8):540–551 (2004). Copyright © 2004, American Society for Blood and Bone Marrow Transplantation.

3.6. The Combination of CD11b-Depleted BM and Low-Dose Donor Splenocytes Led to a Durable GvL Effect without GvHD

We tested whether manipulating the CD11b content of the BM graft, in the absence of added splenocytes, could confer protection against LBRM in the

C57BL/6 → B10.BR model. Administration of LBRM to recipients of unmanipulated BM, CD11b [+] BM, CD11b [+/–] or CD11b [–] BM without donor splenocytes led to uniform death from progressive leukemia by day +80 post-transplant (Figure 3B). Surprisingly, the addition of a low dose of donor splenocytes (0.5×10^6 or 1.0×10^6) to either CD11b [–] BM or CD11b [+/–] BM conferred significant protection against LBRM compared to recipients of the same numbers of donor splenocytes combined with unmanipulated BM ($p = 0.02$ and $p = 0.0002$, respectively) and CD11b [+] BM ($p = 0.0146$ and 0.0005, respectively). Of note, long-term (>200 days) leukemia-free survival was seen in 40% of mice that received 1×10^6 donor splenocytes in combination with CD11b [+/–] BM grafts (Figure 3C). No long-term survival was seen among recipients of CD11b [–] BM and 1.0×10^6 splenocytes that were challenged with LBRM on day –1 due to severe GvHD pathology (data not shown).

Transplantation of a CD11b-depleted BM graft in combination with low-dose splenocytes also conferred resistance to delayed post-transplant challenge with LBRM (day +75), a model that may reflect late leukemia relapse after BMT. Recipients transplanted with unmanipulated BM, CD11b [+] BM, CD11b [+/–] BM, or CD11b [–] BM in the absence of low-dose donor lymphocytes died from progressive leukemia on day +120 to day +155 post-transplantation following LBRM challenge on day +75 (45–80 days after LBRM infusion) (Figure 3D). Complete or partial CD11b-depletion of the BM graft significantly enhanced the survival of transplant recipients compared to recipients of unmanipulated BM grafts ($p < 0.0001$ and $p = 0.03$, respectively) or CD11b [+] BM ($p < 0.0001$ and $p = 0.02$, respectively), with 40% of recipients of CD11b [–] BM showing long-term (>200 days) survival without clinical or histopathological evidence of GvHD or persistent leukemia (Figure 3E).

3.7.   Recipients of CD11b-Depleted BM Grafts Had Increased Numbers of Donor Spleen-Derived Memory T Cells in the Blood Post-Transplant

Since CD11b-depletion of the BM graft was associated with an enhanced graft-versus-leukemia effect (Figure 3C,E), we next studied whether manipulation of the BM graft affected the kinetics of donor T cell reconstitution after BMT. Peripheral blood samples from recipients of unmanipulated BM, CD11b [+] BM, CD11b [+/–] BM, or CD11b [–] BM were analyzed for the presence of circulating donor T cells at days +30, +60, and +105 post-transplant. Donor BM-derived T cells with the phenotype CD45.1$^+$CD90.2$^+$ increased in frequency from 6–10% at day +30 to 14–18% at day +60, and then remained stable as a 16–20% fraction of blood leukocytes (1300–1700 T cells/µL) (Figure 4A). Manipulation of the CD11b content of the BM graft had no effect on overall numbers of donor-BM T cells in the blood. In contrast, depleting CD11b$^+$ cells from the BM graft resulted in a marked expansion of donor spleen-derived T cells with the phenotype CD45.1$^+$CD90.1$^+$ (Figure 4B). There were no significant differences in the

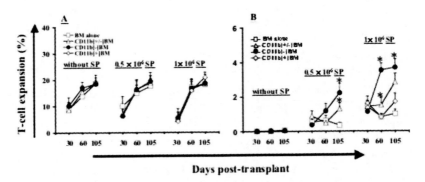

Days post-transplant

Figure 4. Expansion of donor-derived T cells was enhanced by CD11b depletion of the BM graft. B10.BR mice were lethally irradiated with 11 Gy on day –2 and transplanted with 5 x 10^6 unmanipulated C57BL/6 BM (open squares), CD11b[+] BM (open diamonds), CD11b[+/–] BM (open triangles) or CD11b[–] BM (filled circles) plus various numbers of splenocytes. Flow cytometry for T cell analysis was performed on samples of peripheral blood obtained on days +30, +60, and +105 after BMT. The transplant experiment was repeated at least three times for each condition, and 5–10 mice were transplanted for each condition in each experiment. (A) donor BM-derived T cells; (B) donor spleen graft-derived T cells. *$p < 0.05$, **$p < 0.01$, ***$p < 0.001$ compared to the frequencies of spleen-derived T cells in the blood of recipients of unmanipulated BM/splenocytes at same time point.

frequencies of BM-derived naive and memory CD4 or CD8 T cells between recipients of various BM grafts (data not shown). The effect of CD11b-depletion on expansion of spleen-derived T cells was inversely proportional to the content of CD11b$^+$ cells in the BM graft, with mean values of 721 ± 126, 329 ± 79, and 186 ± 46 donor spleen-derived T cells/μL at day +105 post-transplant among recipients of CD11b [–] BM, CD11b [+/–] BM, and unmanipulated BM grafts, respectively. Transplantation of 0.5 × 10^6 spleen cells with CD11b-depleted BM grafts produced the same effect on expansion of spleen-derived donor T cells as was seen among recipients of 1 × 10^6 splenocytes, although the absolute number of spleen-derived T cells in the blood was approximately half (Figure 4B). Examination of the CD4$^+$ T cell subsets and the phenotypically defined naive T cells (CD62L$^+$CD44^{low}) and memory T cell compartments (CD62L$^-$CD44^{high}) revealed that the effect of CD11b depletion of the BM graft on donor T cell expansion was predominately due to expansion of spleen-derived CD4$^+$ memory T cells ($p < 0.01$, data not shown). There was also an increase in the ratio of in CD4:CD8 donor spleen-derived memory T cells among recipients transplanted with splenocytes combined with CD11b [–] BM ($p < 0.05$)[19]. However, there was no effect of CD11b–depletion on the frequency of spleen-derived CD8$^+$ T cells or the spleen-derived CD8$^+$ memory T cell subset.

3.8. Recipients of CD11b-Depleted Allogeneic BMT Had Increased Levels of Serum IFN-γ at Day +30 Post-BMT

In order to explore potential mechanisms for the observed increase in mature donor-derived CD4$^+$ memory T cells following CD11b–depletion of the BM graft, serum levels of inflammatory cytokines were measured at day +30 post-transplantation. Day +30 serum levels of IFN-γ were significantly increased in the recipients transplanted with the combination of 5×10^6 CD11b [–] BM and 0.2×10^6 MACs purified donor splenic T cells, compared to recipients of unmanipulated BM and CD11b [+] BM and the same doses of donor splenic T cells ($p < 0.05$; Figure 5). There was no significant effect of CD11b-depletion on day+30 serum levels of TNF-α (data not shown).

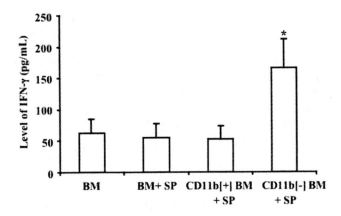

Figure 5. CD11b depletion of BM led to enhanced serum levels of IFN-γ. Serum samples were obtained on day +30 post-transplant. Mean values (±SD) of IFN-γ were determined by ELISA from 5 mice per group. (*$p < 0.05$ comparing IFN-γ levels from the group receiving CD11b-depleted BM to the group that received unmanipulated BM). Reprinted with permission from J.M. Li and E.K. Waller. Donor antigen-presenting cells regulate T-cell expansion and antitumor activity after allogeneic bone marrow transplantation. *Biol Blood Marrow Transplant* **10**(8):540–551 (2004). Copyright © 2004, American Society for Blood and Bone Marrow Transplantation.

4. DISCUSSION

A growing body of data[10,32,33], the data presented here, and unpublished data from our laboratory using FACS purified CD11b$^+$ DC and CD11b$^-$ DC subsets support the concept that DC and DC precursors in resting physiologic conditions may polarize the development of cellular immune responses and prime naive T cell activation and proliferation in response to antigen. We have previously observed that the content of donor plasmacytoid DC precursors in human BM allografts was associated with a higher incidence of post-transplant relapse, and inversely associated with the incidence of chronic GvHD[3]. These and other data have been interpreted to suggest that indirect antigen presentation by donor plasmacytoid DC may result in tolerance of donor T cells to allo-antigen.[34] Thus, we hypothesized that phenotypically defined DC and DC precursors present in a BM graft may inhibit cellular immune responses in the first days to weeks post-transplant. The present study and our recent publication[19] support the role of CD11b[+] donor cells in directing immune reconstitution following allogeneic BMT.

In traditional Chinese philosophy, all phenomena can be described as resulting from the interaction of opposites that simultaneously embrace and oppose each other called "Yin" and "Yang." Elements that are characterized by calm/rest are classified as Yin, and elements that are characterized as being active are classified as Yang. Normal immune response system can be viewed as a manifestation of the balance between tolerance and anergy (Yin) and activation and proliferation (Yang). The optimal condition for a successful allogeneic transplant is coexistence of Yin and Yang with a strong graft-versus-tumor response (Yang) while maintaining tolerance to normal host tissues (Yin). In our work we provide evidence that the content of donor CD11b[+] cells from the BM grafts is responsible, in part, for regulating the balance between too much or too little donor immunity, and that donor APC in the marrow graft can either augment (Yang) or inhibit (Yin) cellular immune responses post-transplantation (Figure 1).

Myeloablative conditioning and transplantation of allogeneic HPC lead to a condition that can be described as representing a deficiency of both Yin and Yang. Early on, transplant recipients have a condition of dysregulated immunity in which effective antigen-specific immune responses are often lacking (deficiency of Yang) and in which uncontrolled graft-versus-host effects can lead to lethal damage to normal tissue (deficiency of Yin). Our study demonstrates that the mechanical manipulation of a non-T cell and non-stem cell compartment of an allogeneic BM graft can enhance post-transplant cellular immune reconstitution in a manner that more appropriately balances immunoregulatory elements (Figure 1) and results in a GvL effect without lethal GvHD. A striking and unexpected finding of the present study is that depletion of a fraction of CD11b[+] donor BM cells led to marked post-transplant expansion of donor T cells associated with increased levels of IFN-γ and enhanced GvL effects, consistent with enhancement of cellular immunity (Yang). Since we have used different congenic strains as BM donors and donors of mature (splenic) T cells, these results demonstrate a trans-effect of regulatory BM cells on donor T cell activation and expansion that is not explained by alterations in the content of donor T cells, or T cell precursors, including pluripotent or committed lymphoid progenitors[35,36]. Recent preliminary data using recipients transplanted with FACS purified CD11b[+] DC and CD11b[-] DC, plus FACS purified HSC and MACs purified splenic T cells, have confirmed these results (Li and Waller, in preparation). These data support our hypothesis that CD11b-depletion removes a regulatory

BM population, adjusting the balance of donor immune regulatory cells and resulting in allo-activation and expansion of donor T cells that mediate a GvL effect.

Another significant observation of the present study is that one of the main effects of varying the content of donor CD11b[+] cells appears to be on the proliferation and accumulation of donor-derived CD4[+] cells without an effect on donor BM-derived T cells[19] (Figure 4). These data suggest that mature T cells derived from spleen cell inoculums rather than de novo naive donor T cells generated during thymopoiesis from HPC contained in the BM graft mediates the major GvL effect. These data further suggest the following mechanistic hypotheses: that the improved GvL activity of donor splenic T cells is associated with (1) donor CD4[+] T cell[22] activation by the CD11b[-] DC present in the BM graft following depletion of CD11b[+] cells, or (2) release of donor T cells from the immune-suppression mediated by donor BM CD11b[+] cell populations. The role of CD4[+] donor T cells in GvL activity is supported by Nimer, who reported that allogeneic donor grafts depleted of CD8[+] cells led to less GvHD than unmanipulated grafts, without a marked increase in relapse rates[37]. Furthermore, Alyea et al. and Giralt et al. reported that depletion of CD8[+] cells from donor leukocyte products resulted in complete remission for CML patients who relapsed after allogeneic BMT with minimal GvHD[38,39]. CD4[+] cell lines harvested from patients post-transplant had lytic activity against CML cells[40]. Our results were consistent with these data and suggest that CD4[+] T cells may mediate GvL, at least in the LBRM model system. Additional studies, now in progress, will distinguish among the possible mechanisms we have proposed.

Of note, recipients of CD11b [−] BM challenged with LBRM on day -1 had more GvHD-related death compared to recipients of CD11b [+/−] BM, and recipients of CD11b [−] BM that were infused with LBRM on day +75 (Figure 3C,E). These data indicate that the larger numbers of CD11b[-]APC in CD11b [−] BM may augment donor T cell activation to the extent that lethal GVHD ensues when large numbers of MHC class-II[+] tumor cells are presented, and in the presence of inflammatory cytokines induced by myeloablative irradiation[41]. The absence of lethal GvHD in recipients of LBRM plus CD11b [+/−](partially depleted) BM is hypothesized to be due to lower levels of CD11b[-] donor APC compared to grafts that were CD11b[−] (fully depleted) BM. The lower rates of GvHD seen when LBRM was administered day +75 following transplantation of CD11b[−] BM likely results from either (1) lower levels of inflammatory cytokines at day +75 compared to day −1[41], or (2) a reduced content of donor CD11b[-] APC following donor hematopoietic reconstitution by day +75 post-transplant, leading to less allo-activation of donor T cells. Thus, these data suggest that the effect of allo-activation of donor T cells that is due to manipulation of the content of BM CD11b[+] cells may vary according to the time at which host-type tumor cells are administered post-transplant.

The proposed model leads to testable hypotheses regarding the role of purified donor DC populations co-transplanted with purified donor stem cells and T cells; these experiments are in progress. Clinical translation of this concept would require a method of selectively removing a human DC subpopulation from an allogeneic BM or blood HPC graft. Commercially available monoclonal antibodies, such as BDCA2, could achieve this without a substantial effect on the content of donor T cells or stem cells. In conclusion, the data presented herein support the hypothesis that predominantly immature donor CD11b$^+$ DC cells in the BM allograft have a significant immunoregulatory effect on post-transplant cellular immune reconstitution in MHC mismatched allogeneic transplantation. Removal of donor CD11b$^+$ cells, an immune suppressive element (Yin), augmented donor T cell activation and proliferation (Yang), resulting in enhanced GvL activity without a marked effect on GvHD in most transplants. Future clinical trials are necessary to determine whether this paradigm can be applied to transplants in humans, and whether manipulation of the analogous human donor dendritic cell subsets in allogeneic HPCT will result in improved leukemia-free survival.

5. ACKNOWLEDGMENTS

The authors are grateful to Dr. Cindy Giver for careful reading of the manuscript and Ms. Marcie Burnham for professional editorial assistance.

6. REFERENCES

1. W.D. Shlomchik, M.S. Couzens, C.B. Tang, J. McNiff, M.E. Robert, J. Liu, M.J. Shlomchik and S.G. Emerson. Prevention of graft versus host disease by inactivation of host antigen-presenting cells. *Science* **285**(5426):412–415 (1999).
2. C.C. Matte, J. Liu, J. Cormier, B.E. Anderson, I. Athanasiadis, D. Jain, J. McNiff and W.D. Shlomchik. Donor APCs are required for maximal GVHD but not for GVL. *Nat Med* **10**(9):987–992 (2004).
3. E.K. Waller, H. Rosenthal, T.W. Jones, J. Peel, S. Lonial, A. Langston, I. Redei, I. Jurickova and M.W. Boyer. Larger numbers of CD4(bright) dendritic cells in donor bone marrow are associated with increased relapse after allogeneic bone marrow transplantation. *Blood* **97**(10):2948–2956 (2001).
4. V. Reddy, J.A. Iturraspe, A.C. Tzolas, H.U. Meier-Kriesche, J. Schold and J.R. Wingard. Low dendritic cell count after allogeneic hematopoietic stem cell transplantation predicts relapse, death and acute graft-versus-host disease. *Blood* **103**(11):4330–4335 (2004).
5. B. Pulendran, J.L. Smith, G. Caspary, K. Brasel, D Pettit, E. Maraskovsky and C.R. Maliszewski. Distinct dendritic cell subsets differentially regulate the class of immune response in vivo. *Proc Natl Acad Sci USA* **96**(3):1036–1041 (1999).
6. R. Maldonado-Lopez, T. De Smedt, P. Michel, J. Godfroid, B. Pajak, C. Heirman, K. Thielemans, O. Leo, J. Urbain and M. Moser. CD8alpha+ and CD8alpha− subclasses of dendritic cells direct the development of distinct T helper cells in vivo. *J Exp Med* **189**(3):587–592 (1999).

7. M.C. Rissoan, V. Soumelis, N. Kadowaki, G. Grouard, F. Briere, R. de Waal Male-fyt and Y.J. Liu. Reciprocal control of T helper cell and dendritic cell differentia-tion. *Science* **283**(5405):1183–1186 (1999).

8. F.P. Siegal, N. Kadowaki, M. Shodell, P.A. Fitzgerald-Bocarsly, K. Shah, S. Ho, S. Antonenko and Y.J. Liu. The nature of the principal type 1 interferon-producing cells in human blood. *Science* **284**(5421):1835–1837 (1999).

9. M. Cella, D. Jarrossay, F. Facchetti, O. Alebardi, H. Nakajima, A. Lanzavecchia and M. Colonna. Plasmacytoid monocytes migrate to inflamed lymph nodes and produce large amounts of type I interferon. *Nat Med* **5**(8):919–923 (1999).

10. F. Sallusto and A. Lanzavecchia. Mobilizing dendritic cells for tolerance, priming, and chronic inflammation. *J Exp Med* **189**(4):611–614 (1999).

11. A. Lanzavecchia and F. Sallusto. The instructive role of dendritic cells on T cell responses: lineages, plasticity and kinetics. *Curr Opin Immunol* **13**(3):291–298 (2001).

12. M.F. Lipscomb and B.J. Masten. Dendritic cells: immune regulators in health and disease. *Physiol Rev* **82**(1):97–130 (2002).

13. M. O'Keeffe, H. Hochrein, D. Vremec, I. Caminschi, J.L. Miller, E.M. Anders, L. Wu, M.H. Lahoud, S. Henri, B. Scott, P. Hertzog, L. Tatarczuch and K. Shortman. Mouse plasmacytoid cells: long-lived cells, heterogeneous in surface phenotype and function, that differentiate into CD8(+) dendritic cells only after microbial stimulus. *J Exp Med* **196**(10):1307–1319 (2002).

14. M. O'Keeffe, H. Hochrein, D. Vremec, B. Scott, P. Hertzog, L. Tatarczuch and K. Shortman. Dendritic cell precursor populations of mouse blood: identification of the murine homologues of human blood plasmacytoid pre-DC2 and CD11c+ DC1 precursors. *Blood* **101**(4):1453–1459 (2003).

15. R. Pelayo, J. Hirose, J. Huang, K.P. Garrett, A. Delogu, M. Busslinger and P.W. Kincade. Derivation of 2 categories of plasmacytoid dendritic cells in murine bone marrow. *Blood* **105**(11):4407–4415 (2005).

16. C. Ruedl and M.F. Bachmann. CTL priming by CD8(+) and CD8(-) dendritic cells in vivo. *Eur J Immunol* **29**(11):3762–3767 (1999).

17. G. Schlecht, C. Leclerc and G. Dadaglio. Induction of CTL and nonpolarized Th cell responses by CD8alpha(+) and CD8alpha(−) dendritic cells. *J Immunol* **167**(8):4215–4221 (2001).

18. A.D. Billiau, S. Fevery, O. Rutgeerts, W. Landuyt and M. Waer. Transient expan-sion of Mac1+Ly6-G+Ly6-C+ early myeloid cells with suppressor activity in spleens of murine radiation marrow chimeras: possible implications for the graft-versus-host and graft-versus-leukemia reactivity of donor lymphocyte infusions. *Blood* **102**(2):740–748 (2003).

19. J.M. Li and E.K. Waller. Donor antigen-presenting cells regulate T-cell expansion and antitumor activity after allogeneic bone marrow transplantation. *Biol Blood Marrow Transplant* **10**(8):540–551 (2004).

20. S. Gillis and S.B. Mizel. T-cell lymphoma model for the analysis of interleukin 1-mediated T-cell activation. *Proc Natl Acad Sci USA* **78**(2):1133–1137 (1981).

21. E.K. Waller, A.M. Ship, S. Mittelstaedt, T.W. Murray, R. Carter, I. Kakhniashvili, S. Lonial, J.T. Holden and M.W. Boyer. Irradiated donor leukocytes promote en-graftment of allogeneic bone marrow in major histocompatibility complex mis-matched recipients without causing graft-versus-host disease. *Blood* **94**(9):3222–3233 (1999).

22. C.R. Giver, R.O. Montes, S. Mittestaedt, J.M. Li, D.L. Jaye, S. Lonial, M.W. Boyer and E.K. Waller. Ex vivo fludarabine exposure inhibits graft-versus-host activity of allogeneic T cells while preserving graft-versus-leukemia effects. *Biol Blood Marrow Transplant* **9**(10):616–632 (2003).

23. D.H. Fowler, K. Kurasawa, A. Husebekk, P.A. Cohen and R.E. Gress. Cells of Th2 cytokine phenotype prevent LPS-induced lethality during murine graft-versus-host reaction: regulation of cytokines and CD8+ lymphoid engraftment. *J Immunol* **152**(3):1004–1013 (1994).

24. R.C. Budd, J.C. Cerottini, C. Horvath, C. Bron, T. Pedrazzini, R.C. Howe and H.R. MacDonald. Distinction of virgin and memory T lymphocytes: stable acquisition of the Pgp-1 glycoprotein concomitant with antigenic stimulation. *J Immunol* **138**(10):3120–3129 (1987).

25. T.L. Walunas, D.S. Bruce, L. Dustin, D.Y. Loh and J.A. Bluestone. Ly-6C is a marker of memory CD8+ T cells. *J Immunol* **155**(4):1873–1883 (1995).

26. J.M. Curtsinger, D.C. Lins and M.F. Mescher. CD8+ memory T cells (CD44high, Ly-6C+) are more sensitive than naive cells to (CD44low, Ly-6C–) to TCR/CD8 signaling in response to antigen. *J Immunol* **160**(7):3236–3243 (1998).

27. K.R. Cooke, L. Kobzik, T.R. Martin, J. Brewer, J. Delmonte Jr., J.M. Crawford and J.L. Ferrara. An experimental model of idiopathic pneumonia syndrome after bone marrow transplantation, I: the roles of minor H antigens and endotoxin. *Blood* **88**(8):3230–3239 (1996).

28. K.R. Cooke, W. Krenger, G. Hill, T.R. Martin, L. Kobzik, J. Brewer, R. Simmons, J.M. Crawford, M.R. van den Brink and J.L. Ferrara. Host reactive donor T cells are associated with lung injury after experimental allogeneic bone marrow transplantation. *Blood* **92**(7):2571–2580 (1998).

29. J.M. Crawford. Graft-versus host disease of the liver. In *Graft-versus-host disease*, pp. 315-333. Ed. J.L. Ferrara, H.J. Deeg and S.J. Burakoff. New York: Marcel Dekker (1996).

30. A. Mowat. Intestinal graft-versus host disease. In *Graft-versus-host disease*, pp. 337–384. Ed. J.L. Ferrara, H.J. Deeg and S.J. Burakoff. New York: Marcel Dekker (1996).

31. D. Stull and S. Gillis. Constitutive production of interleukin 2 activity by a T cell hybridoma. *J Immunol* **126**(5):1680–1683 (1981).

32. C. Kurts, H. Kosaka, F.R. Carbone, J.F. Miller and W.R. Heath. Class I-restricted cross-presentation of exogenous self-antigens leads to deletion of autoreactive CD8(+) T cells. *J Exp Med* **186**(2):239–245 (1997).

33. K. Inaba, S. Turley, F. Yamaide, T. Iyoda, K. Mahnke, M. Inaba, M. Pack, M. Subklewe, B. Sauter, D. Sheff, M. Albert, N. Bhardwaj, I. Mellman and R.M. Steinman. Efficient presentation of phagocytosed cellular fragments on the major histocompatibility complex class II products of dendritic cells. *J Exp Med* **188**(11):2163–2173 (1998).

34. E.K. Waller, H. Rosenthal and L. Sagar. DC2 effect on survival following allogeneic bone marrow transplantation. *Oncology* (Huntington) **16**(1 Suppl 1):19–26 (2002).

35. R.S. Negrin, C.R. Kusnierz-Glaz, B.J. Still, J.R. Schriber, N.J. Chao, G.D. Long, C. Hoyle, W.W. Hu, S.J. Horning, B.W. Brown, K.G. Blume and S. Strober. Transplantation of enriched and purged peripheral blood progenitor cells from a single apheresis product in patients with non-Hodgkin's lymphoma. *Blood* **85**(11):3334–3341 (1995).

36. C. Arber, A. BitMansour, T.E. Sparer, J.P. Higgins, E.S. Mocarski, I.L. Weissman, J.A. Shizuru and J.M. Brown. Common lymphoid progenitors rapidly engraft and protect against lethal murine cytomegalovirus infection after hematopoietic stem cell transplantation. *Blood* **102**(2):421–428 (2003).

37. S.D. Nimer, J. Giorgi, J.L. Gajewski, N. Ku, G.J. Schiller, K. Lee, M. Territo, W. Ho, S. Feig, M. Selch, V. Isacescu, T.A. Reichert and R.E. Champlin. Selective depletion of CD8+ cells for prevention of graft-versus-host disease after bone marrow transplantation: a randomized controlled trial. *Transplantation* **57**(1):82–87 (1994).

38. E.P. Alyea, R.J. Soiffer, C. Canning, D. Neuberg, R. Schlossman, C. Pickett, H. Collins, Y. Wang, K.C. Anderson and J. Ritz. Toxicity and efficacy of defined doses of CD4(+) donor lymphocytes for treatment of relapse after allogeneic bone marrow transplant. *Blood* **91**(10):3671–3680 (1998).

39. S. Giralt, J. Hester, Y. Huh, C. Hirsch-Ginsberg, G. Rondon, D. Seong, M. Lee, J. Gajewski, K. Van Besien, I. Khouri, R. Mehra, D. Przepiorka, M. Korbling, M. Talpaz, H. Kantarjian, H. Fischer, A. Deisseroth and R. Champlin. CD8-depleted donor lymphocyte infusion as treatment for relapsed chronic myelogenous leukemia after allogeneic bone marrow transplantation. *Blood* **86**(11):4337–4343 (1995).

40. K.R. Oettel, O.H. Wesly, M.R. Albertini, J.A. Hank, O. Iliopolis, J.A. Sosman, K. Voelkerding, S.Q. Wu, S.S. Clark and P.M. Sondel. Allogeneic T-cell clones able to selectively destroy Philadelphia chromosome-bearing (Ph1+) human leukemia lines can also recognize Ph1– cells from the same patient. *Blood* **83**(11):3390–3402 (1994).

41. G.R. Hill, J.M. Crawford, K.R. Cooke, Y.S. Brinson, L. Pan and J.L. Ferrara. Total body irradiation and acute graft-versus-host disease: the role of gastrointestinal damage and inflammatory cytokines. *Blood* **90**(8):3204–3213 (1997).

6

IT'S ONLY INNATE IMMUNITY BUT I LIKE IT

Emanuela Marcenaro[1,2], Mariella Della Chiesa[1],
Alessandra Dondero[1], Bruna Ferranti[1],
and Alessandro Moretta[1,3]

1. INTRODUCTION

Natural Killer (NK) cells are capable of discriminating between normal and virus-infected cells or cells undergoing tumor transformation[1,2]. The selective elimination of abnormal target cells, based on classical NK-mediated cytotoxicity, is the result of the combined function of activating receptors including NCR and NKG2D[3,4] and inhibitory receptors such as Killer Ig-like receptors (KIR) and CD94/NKG2A[5–8] on NK cells as well as of the expression of their specific ligands on target cells[9–11].

However, NK cells also display regulatory capabilities mediated by various cytokines released upon engagement of different triggering NK receptors or upon signaling by other cytokines[1-4]. This is particularly relevant during the early phases of inflammatory responses.

Much data have recently highlighted the role of the interactions between NK cells and other cells of the innate immune system that occur during the early phases of acute inflammation, secondary to infection. Various studies were focused on the crosstalk between NK cells and monocyte-derived dendritic cells (MDDC) and more recently on the involvement of plasmacytoid dendritic cells (PDC), mast cells, basophils, eosinophils, and neutrophils. Thus a complicated network of interactions can take place after the recruitment of these different cells to inflammatory sites in response to tissue damage resulting from invasion by pathogens (or tumor cells)[12–24].

[1]Dipartimento di Medicina Sperimentale, Università degli Studi di Genova, Via L.B. Alberti 2, 16132 Italy; [2]Istituto Giannina Gaslini, L.go G. Gaslini 5, 16147 Genova, Italy; [3]Centro di Eccellenza per le Ricerche Biomediche, Università degli Studi di Genova, V.le Benedetto XV, 16132 Genova, Italy. Address all correspondence to: Alessandro Moretta M.D., Dipartimento di Medicina Sperimentale, Sezione di Istologia, Via G.B. Marsano 10, 16132 GENOVA, ITALY. Phone: +39-010-35 37 868. FAX: +39-010-35 37 576. E-mail: alemoret@unige.it

In the present review we discuss how the early crosstalk occurring between various cells of the innate immunity could play an important role in the control of the quality and efficacy of the defenses against pathogens and how these interactions can have an impact on adaptive immune responses.

2. THE IMMUNOREGULATORY ROLE OF NK CELLS: CROSSTALK BETWEEN NK, MDDC, AND PDC

During the early phases of an inflammatory response, NK cells may be recruited in response to various proinflammatory chemokines and cytokines into inflamed tissues and interact with other cell types of the innate immunity[24-26]. These interactions can modulate NK cell functions as the result of mechanisms of cell-to-cell contact (favoring receptor/ligand interactions) and of the activity of soluble factors. For example, a close cell-to-cell contact has been shown to be required during the interactions occurring between activated NK cells and MDDC. However, NK cells recruited from blood into inflamed peripheral tissues may not necessarily be preactivated and would thus require appropriate activating signals in order to interact with autologous MDDC. Such activating signals could be provided by tumors or virus-infected cells susceptible to NK-mediated lysis. Nevertheless, in most instances, tumors are resistant to non-activated peripheral blood NK cells and their killing requires previous exposure of NK cells to cytokines such as IL2, IL12, IL15, or IFN-alpha released by other cell types[1,2,5,6].

Another mechanism by which NK cells can become activated within pathogen-invaded tissues has been identified recently. It has been shown that circulating as well as in-vitro activated human NK cells express Toll-like receptors (TLR)[27-29] and that TLR can provide an alternative mode of NK cell activation, independent on the recognition of NK-susceptible target cells. TLR are pattern recognition receptors (PRR) that, upon recognition of pathogen-associated molecular patterns (PAMP), induce triggering of innate immune responses, providing immediate protection against various pathogens[30]. Human NK cells express functional TLR3 and TLR9 (while they do not seem to express other TLR)[27], thus enabling them to respond both to viral and bacterial products. In particular, the simultaneous engagement of TLR3 on both NK cells and MDDC appears to be sufficient to initiate the early phases of innate immune responses. Indeed, exposure of NK cells to double-stranded (ds) RNA results in NK cell activation, as revealed by the surface expression of CD69 and CD25. In addition, TLR-stimulated NK cells, in the presence of IL12, release cytokines including IFN-gamma and TNF-alpha and acquire a higher cytolytic activity against tumor target cells as well as against immature MDDC.[27] On the other hand, MDDC undergoing maturation after antigen uptake[31-33] release cytokines that can greatly influence the functional behavior of NK cells[19-26]. In particular, MDDC-derived IL12 is crucial for inducing NK cells to release IFN-gamma and also for enhancing NK cell cytotoxicity in response to TLR engagement[24,34].

The "editing process" by which NK cells eliminate immature but not mature MDDC is a most remarkable event occurring during NK–DC crosstalk[19,24]. The NK-mediated killing of MDDC is confined to a phenotypically distinct subset that expresses CD94/NKG2A but not inhibitory KIR[35]. In addition it has been demonstrated that the mechanism of MDDC killing by NK cells involves the NKp30 triggering receptor since antibodies against NKp30 could sharply inhibit this function[15]. Thanks to the editing process, NK cells would prevent the survival of faulty DC, keeping in check the quality of DC undergoing maturation[19,26]. Remarkably, this process, by selecting MDDC characterized by optimal APC capability, has a major impact on the generation of downstream adaptive immune responses characterized by Th1 polarization.

Mailliard and colleagues have recently shown that, in addition to IL12, other cytokines, such as IL18, released by MDDC or by macrophages, in response to pathogens, may influence the "helper" activity of NK cells[36]. They showed that, in contrast to IL2, which selectively promotes the cytotoxic activity of NK cells, IL18 does not enhance the cytolytic activity of NK cells, but induces a distinct "helper" pathway of their differentiation. Indeed IL18, but not other NK cell-activating cytokines, promotes the development of a unique type of helper NK cells characterized by the CD56+/CD83+/CCR7+/CD25+ phenotype. These IL18-induced NK cells appear to display high migratory responsiveness to LN-produced chemokines, a distinctive ability to support IL12 production by MDDC and to promote Th1 responses by CD4+ T cells. It is important to underline that among peripheral NK cells only the minor CD56++, CD16– NK subset expresses CCR7, CXCR3, and upregulated levels of CD62L. As a consequence, only these cells are able to reach lymph nodes, whereas CD56+, CD16+ NK cells that do not express these receptors would not be capable to migrate to secondary lymphoid compartements[19,20]. Indeed these NK cells, which represent >80% of the peripheral blood NK cell pool, are equipped with a different set of chemokine receptors including CXCR1 and CX3CR1 that allow their migration from the blood to inflamed peripheral tissues. The important effect of IL18 might be that of rendering also CD56+, CD16+ NK cells capable of migrating toward secondary lymphoid compartments (SLC), where they could directly influence Th1 polarization.

Another recent study[37] demonstrated that IL18, released by MDDC, might play a role during the early phases of innate immune responses by recruiting plasmacytoid DC (PDC), through IL18-receptor expression. This study also demonstrated that PDC exposed to IL18 skewed the development of adaptive immunity toward Th1 polarization.

Two different studies have recently been focused on the in-vitro interaction between human NK cells and plasmacytoid DC (PDC)[38,39]. In human PDC, the pattern of TLR expression is profoundly different from that of MDDC. Thus PDC do not express TLR1, 2, 3, 4, 5, and 6 but, similar to NK cells, express TLR9, a receptor specific for unmethylated CpG derived from bacteria or from viruses[40]. The abundant release of type I IFN (a potent inducer of NK cell cytotoxicity) by PDC[41], stimulated via TLR9[42], suggests that NK–PDC interaction can result in enhanced antiviral innate protection. Indeed, in the presence of stimulation via TLR9 the NK-PDC interaction resulted in upregulation of the NK-mediated cytotoxicity against various tumor target cells. In turn, NK cells are capable of promoting PDC maturation and of upregulating their production of IFN-alpha in response to CpG[38,39]. It is of note that, while NK cells cannot exert an editing program on PDC due to the poor susceptibility of these cells to NK-mediated lysis, when co-cultured with TLR9-stimulated PDC NK cells acquire lytic activity against immature MDDC. Thus, it is conceivable that cellular interactions occurring between NK and PDC during viral infections may influence the maturation and acquisition of functional competence by by-stander MDDC.

Altogether, the above data suggest that the multidirectional PDC–NK–MDDC crosstalk may deeply affect the outcome of antiviral and anti-tumor immunity by regulating both innate NK cell responses and Th polarization.

3. CROSSTALK BETWEEN INNATE AND ADAPTIVE IMMUNE RESPONSES

The NK-mediated capability of killing virus-infected cells, tumors, or immature MDDC is greatly influenced by the type of cytokines released by bystander cells during innate immune responses. In turn, the apoptotic/necrotic material or heat-shock proteins that are generated as a result of the NK-mediated killing of tumors or virus-infected cells can modulate the function of MDDC or other by-stander innate cells[19,26]. It is likely that, during the early phases of an inflammatory response, the engagement of TLR by PAMP may not be confined to NK and DC but could also involve other cell types, including resident mast cells[43-45], neutrophils[46], eosinophils[47], or plasmacytoid DC (PDC)[48]. As a consequence, these cells, through the release of cytokines other than IL12 (e.g., IL4, IFN-alpha, IL18) could differentially modulate the functional capability of bystander NK and MDDC. In this context, it has recently been proposed that the early exposure of NK cells to IL4 could deviate the subsequent adaptive response toward the acquisition of a non-Th1 phenotype[49]. In particular, while short-term NK cell exposure to IL12 promoted the release of high levels of both IFN-gamma and TNF-alpha and the acquisition of cytolytic activity, exposure to IL4 resulted in poor cytokine production and low cytolytic activity. Remarkably, NK cells exposed to IL12 for a time interval compatible with in-vivo responses, may favor the selection of appropriate mature MDDC for subsequent Th1 cell priming in secondary lymphoid organs. On the contrary, NK cells exposed to IL4 do not exert MDDC selection, and may impair efficient Th1 priming and favor either tolerogenic or Th2-type responses. These data suggest that, depending on the type of cytokines released during the early stages of an inflammatory response, by either resident or recruited cells, NK cells can differentially contribute to the quality and magnitude of innate immune responses (i.e., killing of tumor cells, DC editing, and cytokine release). This, in turn, may deeply impact the type of downstream adaptive immune responses[49,50] (Figure 1).

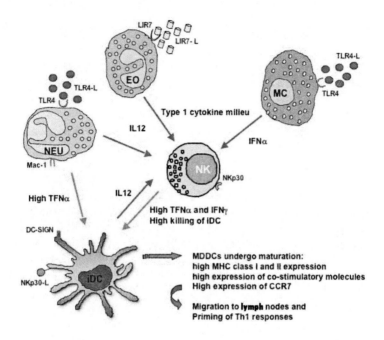

Figure 1. Multidirectional cytokine-based crosstalk between NK and other cells involved in innate immunity. NK cells recruited during acute inflammation can interact with different cell types of the innate immunity that are either resident or have been recruited in response to chemokines produced at the inflammatory sites. These include MDDC, mast cells, eosinophils, and neutrophils. Upon cell activation, induced by stimuli acting on different surface receptors including TLR, each cell type releases different sets of cytokines. These cytokines can greatly influence NK cell function, e.g., their ability to promote MDDC maturation and/or to mediate appropriate editing of MDDC resulting in priming of Th1 cells.

4. INVOLVEMENT OF NEUTROPHILS IN THE REGULATION OF ADAPTIVE IMMUNE RESPONSES THROUGH INTERACTIONS WITH OTHER INNATE EFFECTOR CELLS

As mentioned above, a number of signature structures of pathogens, such as bacterial cell wall components and genetic material from bacteria and viruses, can induce maturation of MDDC through Toll-like receptors (TLR). Moreover, MDDC maturation is supported by direct cell-to-cell interaction with other cell types, including NK cells. In this case the helper effect mediated by NK cells was shown to be TNF-alpha-dependent and promoted by engagement of the NKp30 receptor[51].

Recent studies indicate that activation of MDDC may also involve cellular communication with neutrophils[46,52]. In this context it was proposed that activated neutrophils interact with immature MDDC through binding of DC-SIGN to Lewis moieties on Mac-1 (this is a specific glycoform of Mac-1, that is expressed only on neutrophils)[53]. This molecular interaction appears to be required for neutrophils to induce MDDC maturation. Although the synapse formation through DC-SIGN and Mac-1 does not result in generation of signals promoting direct MDDC maturation, it appears to be required in order to facilitate transmission/delivery of TNF-alpha. Since only activated neutrophils release TNF-alpha[54,55], these cells would favor MDDC maturation only upon encounter with patoghens in inflamed tissues. Neutrophils are the first immune cells to migrate from the blood to sites of infection. They rapidly sense the presence of pathogenic bacteria through TLR, such as TLR2 and TLR4[56], which is important for the immediate killing of these pathogens. Remarkably, TLR-activated neutrophils can also attract other immune cells[57], including other neutrophils through the release of IL8 and GRO-alpha, Th1 cells through MIP-1alpha, MIP-1beta, MIG, IP-10, and I-TAC, as well as immature MDDC through MIP-1alpha, MIP-1beta, and alpha-defensin (see above). In addition, since the CD56+, CD16+ NK cell subset expresses CXCR1, it is likely that the release of IL8 from activated neutrophils may also be involved in the mechanisms of NK cell recruitment at inflammatory sites.

5. OTHER INNATE CELLS SUCH AS MAST CELLS OR EOSINOPHILS ARE IMPORTANT IN THE EARLY PHASES OF INNATE IMMUNE RESPONSES

As mentioned above, the engagement of TLR results in activation of multiple cell types within the innate immune system[30]. These include, for example, NK cells, MDDC, PDC, and neutrophils (see above), but also other innate cells, such as mast cells or eosinophils.

Mast cells are common at sites in the body that are exposed to the external environment — including skin, airways, and gastrointestinal tract — where invading pathogens are frequently encountered. In these locations, mast cells are present in close proximity to blood vessels, where they can regulate vascular permeability and effector cell recruitment[58]. Mast cells can modulate the behavior of these effectors cells through release of mediators. Indeed, they produce a wide variety of cytokines and chemokines, including classical proinflammatory mediators, such as TNF-alpha or cytokines that are associated with antiinflammatory or immunomodulatory effects, such as IL10 and also chemokines, such as CXCL8, that are important for recruitment of other innate cells, including NK cells. Through the selective production of mediators, mast cells can regulate the type of effector cell that is recruited and retained in response to specific types of infection. The ability of mast cells to rapidly release preformed TNF might be of particular importance for generation of responses in lymph nodes. In this context, it has been suggested that mast cells might influence DC migration, maturation, and function via the release of chemokines, such as CCL20, and of proinflammatory cytokines, including abundant TNF-alpha[47]. Remarkably, mast cells produce TNF-alpha in response to TLR4 engagement by LPS[43], suggesting their possible involvement in the maturation process of myeloid DC stimulated as well via TLR4. On the other hand, mast cells when stimulated via TLR3 produce high doses of type I IFN[44]. This suggests their direct involvement in innate antiviral defenses as well as in potentiating the lytic activity of NK cells that are also simultaneously stimulated via TLR3. In addition, triggering of mast cells via TLR2, upon exposure to the Gram-positive bacteria cell wall component peptidoglycan (PGN), favors the release of type 2 cytokines, including IL4, IL5, and IL6[43]. In this case, as mentioned above, the release of IL4 may deviate the subsequent adaptive response toward Th2 by acting at the early stages of the innate immune reaction at inflammatory sites in peripheral tissues. Indeed, IL4, when added to NK cells simultaneously with IL2 or IL12, has also been shown to counteract the effect of these cytokines by suppressing IFN-gamma and TNF-alpha production[49]. In vivo, type 1 and type 2 cytokines are secreted and exert their regulatory role on bystander cells within a short time interval after pathogen invasion.[24] It has been estimated that the window of time available for an innate response to take place is only a few hours, whereas a few days are required for development of primary T cell-mediated specific responses[59,60]. Thus, it appears that, depending on the PAMP present at the site of an infection, mast cells can be activated via different TLR expressed on their surface[61].

Another type of innate cell that is present in peripheral tissue, during inflammatory responses, is represented by eosinophils. These cells are prominent in Th2-driven immune responses, including asthma, allergic, and parasitic disease[62]. However, eosinophils are also recognized as immunomodulatory cells, since they have the ability to interact with T and B lymphocytes and possibly also with NK cells. The crosstalk between eosinophils and lymphocytes can be based on direct molecular interactions with costimulatory surface molecules expressed on eosinophils (e.g., CD28, CD40, CD86, and MHC class II), or via indirect stimulation by mediators secreted by eosinophils.[63] In fact, eosinophils have been demonstrated to contain a multitude of preformed cytokines and chemokines, including Th1 cytokines, such as IFN-gamma and IL12, or Th2 cytokines, such as IL4. Piecemeal degranulation appears to be the major secretory pathway of eosinophils[64]. Distinct signaling cascades may function to selectively mobilize the preformed cytokine proteins that are stored in eosinophil granules. Indeed, it is of note that cytokines with opposing functions, such as IL12 and IL4, may be stored within the same granules, but their release is regulated by distinct stimulatory pathways. For instance, stimulation of eosinophils by the CC chemokines eotaxin or (RANTES)/CC chemokine ligand 5 (CCL5) appears to induce IL4 release[65], whereas stimulation through the engagement of surface molecules such as CD9 or LIR7 triggers IL12 release[66]. These results indicate that eosinophils granule-derived IL4 and IL12 may have mutually exclusive mechanisms of vesicle secretion. The capability of eosinophils not only to rapidly release preformed cytokines but also to differentially regulate which cytokines are released in response to different stimuli suggests that these cells may have an important role both in innate and adaptive immune responses.

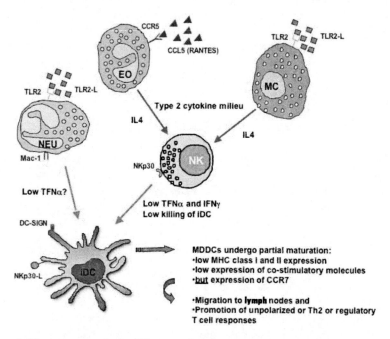

Figure 2. The same actors but a different script: the cytokine milieu during innate immune responses may also deviate priming of T cells toward non-Th1 polarization. The same set of innate cells may be induced to release a different set of cytokines depending on the stimuli acting at the cell surface. For example, mast cells stimulated via TLR2 (recognizing PGN produced by *Staphylococcus aureus*) release IL4 instead of IFN-alpha, whereas eosinophils stimulated by CCL5 (instead of LIR7 ligands) may release IL4 instead of IL12. Remarkably, early exposure of NK cells to IL4 results in downregulation of their responsiveness to IL12 and IL2 and in inhibition of cytolytic activity and cytokine production. As a consequence, IL4-exposed NK cells would favor unpolarized or Th2 as well as tolerogenic responses.

In conclusion, a variety of stimuli can activate mast cells or eosinophils to release a diverse array of biologically active products, many of which can mediate potential proinflammatory, antiinflammatory, and/or immunoregulatory effects. Furthermore, mast cells and eosinophils can be activated differentially to release distinct patterns of mediators or cytokines, depending on the type and strength of the stimuli. This functional and phenotypic plasticity allows their prompt and appropriate response to distinct challenges that are associated with various immunological and pathological responses (Figure 2).

6. CONCLUDING REMARKS

An important concept emerging from a number of recent studies focused on the functional capabilities of cells involved in innate immune responses is their ability to interact with each other via direct cell-to-cell contacts and via the release a large variety of soluble factors. These observations suggest the existence of a remarkable network within different cellular components of the innate immune system[26]. This network can control not only the strength and quality of innate responses but also of the subsequent adaptive immunity. A better understanding of this network and of the interactions among its cellular and molecular components might be of great help in designing new vaccines against infectious agents and, possibly, tumors.

7. ACKNOWLEDGMENTS

This work was supported by grants awarded by the Associazione Italiana per la Ricerca sul Cancro (A.I.R.C.), Istituto Superiore di Sanità (I.S.S.), Ministero della Salute–RF 2002/149, Ministero dell'Istruzione dell'Università e della Ricerca (M.I.U.R.), FIRB-MIUR progetto–cod.RBNE017B4C; Ministero dell' Università e della Ricerca Scientifica e Tecnologica (M.U.R.S.T.), European Union FP6, LSHB-CT-2004-503319-Allostem and Compagnia di San Paolo.

8. REFERENCES

1. G. Trinchieri. Biology of natural killer cells. *Adv Immunol* **47**:187–376 (1989).
2. A. Moretta, C. Bottino, M.C. Mingari, R. Biassoni and L. Moretta. What is a natural killer cell? *Nat Immunol* **3**(1):6–8 (2002).
3. A. Moretta, C. Bottino, M. Vitale, D. Pende, C. Cantoni, M.C. Mingari, R.Biassoni and L. Moretta. Activating receptors and coreceptors involved in human natural killer cell-mediated cytolysis. *Annu Rev Immunol* **19**:197–223 (2001).
4. E. Vivier, J.A. Nunes and F. Vely. Natural killer cell signaling pathways. *Science* **306**(5701):1517–1519 (2004).
5. A. Moretta, C. Bottino, M.Vitale, D. Pende, R. Biassoni, M.C. Mingari and L. Moretta. Receptors for HLA class-I molecules in human natural killer cells. *Annu Rev Immunol* **14**:619–648 (1996).
6. E.O. Long. Regulation of immune responses through inhibitory receptors. *Annu Rev Immunol* **17**:875–904 (1999).
7. M. Lopez-Botet, M. Llano, F. Navarro and T. Bellon. NK cell recognition of non-classical HLA class I molecules. *Semin Immunol* **12**(2):109–119 (2000).

8. C. Vilches and P. Parham. KIR: diverse, rapidly evolving receptors of innate and adaptive immunity. *Annu Rev Immunol* **20**:217–251 (2002).

9. C. Bottino, R. Castriconi, L. Moretta and A. Moretta. Cellular ligands of activating NK receptors. *Trends Immunol* **26**(4):221–226 (2005).

10. A. Cerwenka and L.L. Lanier. Ligands for natural killer cell receptors: redundancy or specificity. *Immunol Rev* **181**:158–169 (2001).

11. D.H. Raulet. Roles of the NKG2D immunoreceptor and its ligands. *Nat Rev Immunol* **3**(10):781–790 (2003).

12. N.C. Fernandez, A. Lozier, C. Flament, P. Ricciardi-Castagnoli, D. Bellet, M. Suter, M. Perricaudet, T. Tursz, E. Maraskovsky and L. Zitvogel. Dendritic cells directly trigger NK cell functions: cross-talk relevant in innate anti-tumor immune responses in vivo. *Nat Med* **5**(4):405–411 (1999).

13. J.L. Wilson, L.C. Heffler, J. Charo, A. Scheynius, M.T. Bejarano and H.G. Ljung-gren. Targeting of human dendritic cells by autologous NK cells. *J Immunol* **163**(12):6365–6370 (1999).

14. E. Carbone, G. Terrazzano, G. Ruggiero, D. Zanzi, A. Ottaiano, C. Manzo, K. Karre and S. Zappacosta. Recognition of autologous dendritic cells by human NK cells. *Eur J Immunol* **29**(12):4022–4029 (1999).

15. G. Ferlazzo, M.L. Tsang, L. Moretta, G. Melioli, R. M Steinman and C. Munz. Human dendritic cells activate resting natural killer (NK) cells and are recognized via the NKp30 receptor by activated NK cells. *J Exp Med* **195**(3):343–351 (2002).

16. F. Gerosa, B. Baldani-Guerra, C. Nisii, V. Marchesini, G. Carra and G. Trinchieri. Reciprocal activating interaction between natural killer cells and dendritic cells. *J Exp Med* **195**(3):327–333 (2002).

17. D. Piccioli, S. Sbrana, E. Melandri and N.M. Valiante. Contact-dependent stimulation and inhibition of dendritic cells by natural killer cells. *J Exp Med* **195**(3):335–341 (2002).

18. M. Vitale, M. Della Chiesa, S. Carlomagno, C. Romagnani, A. Thiel, L. Moretta and A. Moretta. The small subset of CD56brightCD16- natural killer cells is selectively responsible for both cell proliferation and interferon-gamma production upon interaction with dendritic cells. *Eur J Immunol* **34**(6):1715–1722 (2004).

19. A. Moretta. Natural killer cells and dendritic cells: rendezvous in abused tissues. *Nat Rev Immunol* **2**(12):957–964 (2002).

20. M.A. Cooper, T.A. Fehniger, A. Fuchs, M. Colonna and M.A. Caligiuri. NK cell and DC interactions. *Trends Immunol* **25**(1):47–52 (2004).

21. L. Zitvogel. Dendritic and natural killer cells cooperate in the control/switch of innate immunity. *J Exp Med* **195**(3):F9–F14 (2002).

22. D.H. Raulet. Interplay of natural killer cells and their receptors with the adaptive immune response. *Nat Immunol* **5**(10):996–1002 (2004).

23. T. Walzer, M. Dalod, S.H. Robbins, L. Zitvogel and E. Vivier. Natural-killer cells and dendritic cells: "l'union fait la force". *Blood* **106**(7):2252–2258 (2005).

24. M. Della Chiesa, S. Sivori, R. Castriconi, E. Marcenaro and A. Moretta. Pathogen-induced private conversations between natural killer and dendritic cells. *Trends Microbiol* **13**(3):128–136 (2005).

25. A. Moretta. The dialogue between human natural killer cells and dendritic cells. *Curr Opin Immunol* **17**(3):306–311 (2005).

26. A. Moretta, E. Marcenaro, S. Sivori, M. Della Chiesa, M. Vitale and L. Moretta. Early liaisons between cells of the innate immune system in inflamed peripheral tissues. *Trends Immunol* **26**(12):668–675 (2005).

27. S. Sivori, M. Falco, M. Della Chiesa, S. Carlomagno, M. Vitale, L. Moretta and A. Moretta. CpG and double-stranded RNA trigger human NK cells by Toll-like receptors: induction of cytokine release and cytotoxicity against tumors and dendritic cells. *Proc Natl Acad Sci USA* **101**(27):10116–10121 (2004).

28. K.N. Schmidt, B. Leung, M. Kwong, K.A. Zarember, S. Satyal, T.A. Navas, F. Wang and P.J. Godowski. APC-independent activation of NK cells by the Toll-like receptor 3 agonist double-stranded RNA. *J Immunol* **172**(1):138–143 (2004).

29. S. Pisegna, G. Pirozzi, M. Piccoli, L. Frati, A. Santoni and G. Palmieri. p38 MAPK activation controls the TLR3-mediated up-regulation of cytotoxicity and cytokine production in human NK cells. *Blood* **104**(13):4157–4164 (2004).

30. S. Akira and K. Takeda. Toll-like receptor signalling. *Nat Rev Immunol* **4**(7):499–511 (2004).

31. F. Sallusto and A. Lanzavecchia. Mobilizing dendritic cells for tolerance. priming and chronic inflammation. *J Exp Med* **189**(4):611–614 (1999).

32. J. Banchereau, F. Briere, C. Caux, J. Davoust, S. Lebecque, Y.J. Liu, B. Pulendran and K. Palucka. Immunobiology of dendritic cells. *Annu Rev Immunol* **18**:767–811 (2000).

33. C. Reis e Sousa. Dendritic cells as sensors of infection. *Immunity* **14**(5):495–498 (2001).

34. G. Trinchieri. Interleukin-12 and the regulation of innate resistance and adaptive immunity. *Nat Rev Immunol* **3**(2):133–146 (2003).

35. M. Della Chiesa, M. Vitale, S. Carlomagno, G. Ferlazzo, L. Moretta and A. Moretta. The natural killer cell-mediated killing of autologous dendritic cells is confined to a cell subset expressing CD94/NKG2A, but lacking inhibitory killer Ig-like receptors. *Eur J Immunol* **33**(6):1657–1666 (2003).

36. R.B. Mailliard, S.M. Alber, H. Shen, S.C. Watkins, J.M. Kirkwood, R.B. Herberman and P. Kailnski. IL-18-induced CD83+ CCR7+ NK helper cells. *J Exp Med* **200**(7):941–953 (2005).

37. A. Kaser, S. Kaser, N.C. Kaneider, B. Enrich, C.J. Wiedermann and H. Tilg. Interleukin-18 attracts plasmacytoid dendritic cells (DC2s) and promotes Th1 induction by DC2 through IL-18 receptor expression. *Blood* **103**(2):648–655 (2004).

38. F. Gerosa, A. Gobbi, P. Zorzi, S. Burg, F. Briere, G. Carra and G. Trinchieri. The reciprocal interaction of NK cells with plasmacytoid or myeloid dendritic cells profoundly affects innate resistance functions. *J Immunol* **174**(2):727–734 (2005).

39. C. Romagnani, M. Della Chiesa, S. Kohler, B. Moewes, A. Radbruch, L. Moretta, A. Moretta and A. Thiel. Activation of human NK cells by plasmacytoid dendritic cells and its modulation by CD4+ T helper cells and CD4+ CD25hi T regulatory cells. *Eur J Immunol* **35**(8):2452–2458 (2005).

40. D. Jarrossay, G. Napolitani, M. Colonna, F. Sallusto and A. Lanzavecchia. Specialization and complementarity in microbial molecule recognition by human myeloid and plasmacytoid dendritic cells. *Eur J Immunol* **31**(11):3388–3393 (2001).

41. C. Asselin-Paturel and G. Trinchieri. Production of type I interferons: plasmacytoid dendritic cells and beyond. *J Exp Med* **202**(4):461–465 (2005).

42. A.M. Krieg. CpG motifs in bacterial DNA and their immune effects. *Annu Rev Immunol* **20**:709–760 (2002).

43. V. Supajatura, H. Ushio, A. Nakao, S. Akira, K. Okumura, C. Ra and H. Ogawa. Differential responses of mast cell Toll-like receptors 2 and 4 in allergy and innate immunity. *J Clin Invest* **109**(10):1351–1359 (2002).

44. M. Kulka, L. Alexopoulou, R.A. Flavell and D.D. Metcalfe. Activation of mast cells by double-stranded RNA: evidence for activation through Toll-like receptor 3. *J Allergy Clin Immunol* **114**(1):174–182 (2004).

45. J.S. Marshall. Mast-cell responses to pathogens. *Nat Rev Immunol* **4**(10):787–799 (2004).

46. K.P. van Gisbergen, T.B. Geijtenbeek and Y. van Kooyk. Close encounters of neutrophils and DCs. *Trends Immunol* **26**(12):626–631 (2005).

47. H. Nagase, S. Okugawa, Y. Ota, M. Yamaguchi, H. Tomizawa, K. Matsushima, K. Ohta, K. Yamamoto and K. Hirai. Expression and function of Toll-like receptors in eosinophils: activation by Toll-like receptor 7 ligand. *J Immunol* **171**(8):3977–3982 (2003).

48. M. Colonna, G. Trinchieri and Y.J. Liu. Plasmacytoid dendritic cells in immunity. *Nat Immunol* **5**(12):1219–1226 (2004).

49. E. Marcenaro, M. Della Chiesa, F. Bellora, S. Parolini, R. Millo, L. Moretta and A. Moretta. IL-12 or IL-4 prime human NK cells to mediate functionally divergent interactions with dendritic cells or tumors. *J Immunol* **174**(7):3992–3998 (2005).

50. R.B. Mailliard, Y.I. Son, R. Redlinger, P.T. Coates, A. Giermasz, P.A. Morel, W.J. Storkus and P. Kalinski. Dendritic cells mediate NK cell help for Th1 and CTL responses: two-signal requirement for the induction of NK cell helper function. *J Immunol* **171**(5):2366–2373 (2003).

51. M. Vitale, M. Della Chiesa, S. Carlomagno, D. Pende, M. Arico, L. Moretta and A. Moretta. NK-dependent DC maturation is mediated by TNFalpha and IFNgamma released upon engagement of the NKp30 triggering receptor. *Blood* **106**(2):566–571 (2005).

52. S. Bennouna, S.K. Bliss, T.J. Curiel and E.Y. Denkers. Cross-talk in the innate immune system: neutrophils instruct recruitment and activation of dendritic cells during microbial infection. *J Immunol* **171**(11):6052–6058 (2003).

53. K.P. van Gisbergen, M. Sanchez-Hernandez, T.B. Geijtenbeek and Y. van Kooyk. Neutrophils mediate immune modulation of dendritic cells through glycosylation-dependent interactions between Mac-1 and DC-SIGN. *J Exp Med* **201**(8):1281–1292 (2005).

54. M.A. Cassatella. The production of cytokines by polymorphonuclear neutrophils. *Immunol Today* **16**(1):21–26 (1995).

55. S. Bennouna and E.Y. Denkers. Microbial antigen triggers rapid mobilization of TNF-alpha to the surface of mouse neutrophils transforming them into inducers of high-level dendritic cell TNF-alpha production. *J Immunol* **174**(8):4845–4851 (2005).

56. I. Sabroe, L.R. Prince, E.C. Jones, M.J. Horsburgh, S.J. Foster, S.N. Vogel, S.K. Dower and M.K. Whyte. Selective roles for Toll-like receptor (TLR)2 and TLR4 in the regulation of neutrophil activation and life span. *J Immunol* **170**(10):5268–5275 (2003).

57. P. Scapini, J.A. Lapinet-Vera, S. Gasperini, F. Calzetti, F. Bazzoni and M.A. Cassatella. The neutrophil as a cellular source of chemokines. *Immunol Rev* **177**:195–203 (2000).

58. S.J. Galli, S. Nakae and M. Tsai. Mast cells in the development of adaptive immune responses. *Nat Immunol* **6**(2):135–142 (2005).

59. C.A. Biron, K.B. Nguyen, G.C. Pien, L.P. Cousens and T.P. Salazar-Mather. Natural killer cells in antiviral defense: function and regulation by innate cytokines. *Annu Rev Immunol* **17**:189–220 (1999).

60. S.L. Constant and K. Bottomly. Induction of Th1 and Th2 CD4+ T cell responses: the alternative approaches. *Annu Rev Immunol* **15**:297–322 (1997).

61. E. Marcenaro, B. Ferranti and A. Moretta. NK-DC interaction: on the usefulness of auto-aggression. *Autoimmun Rev* **4**(8):520–525 (2005).

62. D. Dombrowicz and M. Capron. Eosinophils, allergy and parasites. *Curr Opin Immunol* **13**(6):716–720 (2001).

63. C. Bandeira-Melo and P.F. Weller. Mechanisms of eosinophil cytokine release. *Mem Inst Oswaldo Cruz* **100**(Suppl 1)73–81 (2005).

64. B. Lamkhioued, A.S. Gounni, D. Aldebert, E. Delaporte, L. Prin, A. Capron and M. Capron. Synthesis of type 1 (IFN gamma) and type 2 (IL-4, IL-5, and IL-10) cytokines by human eosinophils. *Ann NY Acad Sci* **796**:203–208 (1996).

65. C. Bandeira-Melo, K. Sugiyama, L.J. Woods and P.F. Weller. Cutting edge: eotaxin elicits rapid vesicular transport-mediated release of preformed IL-4 from human eosinophils. *J Immunol* **166**(8):4813–4817 (2001).

66. M. Grewe, W. Czech, A. Morita, T. Werfel, M. Klammer, A. Kapp, T. Ruzicka, E. Schopf and J. Krutmann. Human eosinophils produce biologically active IL-12: implications for control of T cell responses. *J Immunol* **161**(1):415–420 (1998).

7

INNATE TUMOR IMMUNE SURVEILLANCE

Mark J. Smyth, Jeremy Swann, and Yoshihiro Hayakawa

1. INTRODUCTION

The innate immune system is emerging as an important mediator of host response to tumor initiation, growth, and metastases. Several chemical carcinogenesis models in mice have implicated an important role for natural killer (NK) cells, $\gamma\delta^+$ T cells, and a specialized CD1d-restricted population of T cells bearing NK cell receptors, called NKT cells[1,2]. The incidence of both methylcholanthrene (MCA)-induced fibrosarcoma and dimethylbenzanthracene (DMBA) and phorbol ester (TPA)-induced skin carcinomas is higher in mice lacking these vital innate immune cells. Furthermore, in allo-bone marrow transplantation (BMT), HLA class I disparities that induce NK cell alloreactions and GVHD also mediate strong graft-versus-leukemia (GVL) effects, producing higher engraftment rates and protecting patients from GVHD[3,4]. In murine MHC-mismatched transplant models with no donor T cell reactivity against the recipient, the pre-transplant infusion of donor-vs.-recipient alloreactive NK cells obviated the need for high-intensity conditioning and conditioned the recipients to BMT without GVHD[3,4]. Several other studies also strongly support a key role for innate cells and mechanisms in controlling tumor initiation and growth[5-8].

2. TYPE I INTERFERON

The potential antitumor function of endogenously produced interferon (IFN)-α/β and exogenously administered type I IFNs has been recognized for a long time. While it had been postulated that IFN acted primarily through host-dependent mechanisms [either via stimulating NK cell proliferation, TNF-related apoptosis-inducing ligand (TRAIL)-mediated cytotoxicity, cytokine secretion, or

Mark J. Smyth, Jeremy Swann, and Yoshihiro Hayakawa, Cancer Immunology Program, Peter MacCallum Cancer Centre, East Melbourne, 3002, Victoria, Australia.

influencing adaptive-immune responses by upregulating class I and class II MHC, antigen presentation, memory T cell survival, or CD8[+]T cell recruitment], most of these concepts have been formulated by experiments where IFN was exogenously administered. Now, Dunn et al.[8,9] have specifically demonstrated that host IFN-α/β prevents the outgrowth of primary carcinogen (methylcholan-threne (MCA))-induced tumors and that several MCA sarcomas derived from IFNα receptor 1-deficient mice were rejected in a lymphocyte-dependent man-ner when transplanted into wild-type mice. By ectopically restoring IFNαR1 or IFNγR expression of sarcomas from receptor-deficient mice and their sensitivity to host immunity, this study has revealed that these tumor cells are not important targets of endogenously produced type I IFN, but rather, IFN-α/β exerted its anti-tumor actions through its capacity to act on host hematopoietic cells. Our own unpublished data using IFNαR1[-/-] and IFNαR2[-/-] mice on a C57BL/6 back-ground support the findings of Dunn et al.[8] IFN-α/β may play a critical intrinsic role in preventing cellular transformation early in the process and then function subsequently extrinsically at the level of the host hematopoietic cells to promote the development of protective tumor specific immune responses. The identity of the IFN-α/β-producing cells that participate in the host protective anti-tumor immune responses remains unknown. While all cells are capable of producing IFN-α/β in response to viral infection, a specialized subset of plasmacytoid den-dritic cells known as interferon-producing cells (IPC) produce especially high levels of IFN-α/β during viral challenge[10]. IPC produce type I IFN as a result of the interaction of pathogen structures with toll-like receptors (TLR); however, thus far no study has clearly demonstrated a role for this subset or TLR in tumor immune surveillance. Our studies thus far have not illustrated a unique role for IPC in tumor control.

3. NKG2D

NKG2D is a key homodimeric activation receptor expressed on the cell surface of almost all NK cells, γδ cells, some cytolytic CD8[+] αβ T cells and NKT cells, and a small subset of CD4[+] αβ T cells[11-14]. Several ligands that bind to NKG2D are members of the MHC class Ib family[14,15]. In humans, the polymorphic MHC class I chain-related molecules (MIC) A and MICB can be recognized by NKG2D[12,16]. Although MIC molecules have not been found in mice, the retinoic acid early inducible-1 (Rae-1) gene products, UL16-binding protein-like tran-script 1 (Mult1), and a distantly related minor histocompatibility Ag, H60, have been reported as NKG2D ligands in mice[14,15,17,18].

MHC Class Ib gene products, including NKG2D ligands, are upregulated by heat shock, retinoic acid, IFN-γ, TLR signaling, growth factors, viral infec-tion, DNA damage, and UV irradiation[19-23]. More recently, a DNA damage re-sponse pathway initiated by ataxia telangiectasia, mutated (ATM), has been shown to regulate expression of NKG2D ligands [24]. In the mouse, the expression of the MHC Class Ib molecules, Rae-1 and H60, is negligible in normal skin, but is strongly induced by skin painting with chemical carcinogens[2]. Natural or induced expression of NKG2D ligands markedly enhances the sensitivity of tumor cells to NK cells in vitro[12–14,17,25,26]. Expression of NKG2D ligands by tumor cells also results in immune destruction in vivo and the ectopic expression of NKG2D ligands, Rae-1 and H60, in several tumor cell lines results in rejection of the tumor cells expressing normal levels of MHC class I molecules[15,27]. Immune-depletion and other studies showed that rejection was dependent on NK cells and/or CD8[+] T cells and perforin[27,28]. While previous studies have illustrated the importance of this pathway in the host immune response to subcutaneous or intravenous injection of experimental tumors either naturally or ectopically expressing various NKG2D ligands, these studies tell us nothing about the role of NKG2D in tumorigenesis[15,27–30]. Despite the assumption that NKG2D-mediated engagement of stress-induced ligands may be a key aspect of tumor immune surveillance[2,31], until recently no study had ever evaluated the importance of the pathway in de-novo tumorigenesis. We have now shown the importance of the NKG2D activation receptor in controlling the natural and activated host response to spontaneous malignancy[32]. Neutralization of NKG2D enhances the sensitivity of wild-type C57BL/6 and BALB/c mice to methylcholanthrene (MCA)-induced fibrosarcoma. The importance of the NKG2D pathway was additionally illustrated in mice deficient for either IFN-γ or TRAIL, while mice depleted of NK cells, T cells, or deficient for perforin did not display any detectable NKG2D phenotype. Furthermore, IL-12 therapy preventing MCA-induced sarcoma formation was also largely dependent upon the NKG2D pathway. While NKG2D ligand expression was variable or absent on sarcomas emerging in wild-type mice, sarcomas derived from perforin-deficient mice or mice neutralized for NKG2D were universally Rae-1[+] and immunogenic when transferred into wild-type syngeneic mice. These findings suggest an important early role for NKG2D in controlling and shaping tumor formation. Experiments performed using the MCA-induced sarcoma model suggest that NK, invariant NKT, $\gamma\delta^+$T cells, and other T cells[1,33–36] may all play a role in host response to the initiation of these tumors. Of note, NKG2D neutralization did not further sensitize WT mice depleted of NK cells, perforin-deficient mice, or RAG-1[−/−] mice deficient in both T and B cells. Clearly, NKG2D ligand expression by tumor cells may not be a barrier to tumor growth since many primary tumors and tumor cell lines naturally express NKG2D ligands, and in some cases these ligands may be secreted[37,38]. Direct experimentation has shown that less tumor rejection occurred when tumors only expressed intermediate levels of Rae-1[27], so it may be that a threshold level of ligand expression can be maintained on a number of tumor cells in the population to effectively downregulate immunity. Recently it has been shown that sustained local expression of NKG2D ligands can impair natural cytotoxicity and tumor immune surveillance[39]. It now remains to be determined whether NKG2D plays an important role in other models of tumor immune surveillance and whether at later stages of tumor development there is some survival advantage for some tumors to express NKG2D ligands.

NKG2D is an important receptor in the recognition of target cells by NK cells, but not the only one. Indeed, some target cells that lack expression of NKG2D ligands are nevertheless sensitive to NK cells, in line with the identification of other NK cell stimulatory receptors that participate in tumor cell recognition[40]. Definition of more of these non-NKG2D-mediated pathways will now be an important step forward.

4. CYTOKINES THAT ACT VIA NKG2D

Cytokines have played an important role in the new progress in tumor immunology and immunotherapy[41]. Other promising cytokines in cancer immunotherapy — including IL-12[42], IL-21[43], IL-18[44], or combination IL-2/IL-18[45] — have also been shown to mediate their anti-tumor activities in mice to a large extent via NK cells. IL-21 has potent anti-tumor activity in a variety of mouse tumor models[43,46-48]. In part, it appears that the anti-tumor activity of IL-21 can be attributed to its ability to induce terminal differentiation of NK cells[43] and regulate T cell proliferation and differentiation[49,48]. We have now illustrated using a series of cytokines with distinct means of activating NK cell effector function, that the NKG2D-NKG2D ligand recognition pathway is pivotal in the anti-metastatic activity of cytokines that promote perforin-mediated cytotoxicity[29,50]. In particular, IL-2, IL-12, or IL-21 suppressed tumor metastases largely via NKG2D ligand recognition and perforin-mediated cytotoxicity. By contrast, IL-18 required tumor sensitivity to FasL, and surprisingly did not depend upon the NKG2D–NKG2D ligand pathway. IL-21 has also been used in combination with IL-2, IL-12, IL-15, and the CD1d reactive glycolipid, α-galactosylceramide (α-GalCer), and great synergy in anti-tumor activity has been observed for IL-15/IL-21 and α-GalCer/IL-21 combinations[51,52].

These studies now provide a fundamental basis for some rational selection of cytokines in NK cell-mediated therapy of tumor metastases that either have or lack NKG2D ligand expression. Given our experimental data, it will be critical to retrospectively assess tumor NKG2D ligand expression and secretion in the large number of non-responders and responders that have taken part in previous clinical trials of cytokine therapies.

5. ACKNOWLEDGMENTS

The authors acknowledge Program Grant and Research Fellowship support from the National Health and Medical Research Council of Australia (M.J.S.), an Australian Postgraduate Award to J.S. (Department of Pathology, University of Melbourne), and a Postdoctoral Fellowship from the Cancer Research Institute (Y.H.).

Table 1. Innate Immunity to Tumors

Mice or treatment	Deficiency	Tumor susceptibility
TCRδ$^{-/-}$	γδ T cells	MCA-induced sarcomas DMBA/TPA-induced skin tumors [2]
Jα18$^{-/-}$	Vα14 TCR$^+$ NKT cells	MCA-induced sarcomas[1,34]
Anti-asialoGM1	NK cells	MCA-induced sarcomas[34]
Anti-NK1.1	NK cells	MCA-induced sarcomas[1,34]
IFNγR1$^{-/-}$	IFN-γ receptor 1, IFN-γ sensitivity	MCA-induced sarcomas Wider spectrum in p53$^{-/-}$ background[53]
IFN-γ$^{-/-}$	IFN-γ production	MCA-induced sarcomas[54] Spontaneous disseminated lymphomas Spontaneous lung adenocarcinomas in BALB/c background[55] HLTV1-tax induced T cell lymphoma[56]
Anti-IFN-α/β	IFN-α/β	Sygeneic tumors[57]
IFNαR1$^{-/-}$	Type I IFN	MCA-induced sarcomas[8] responsiveness
STAT1$^{-/-}$	IFN-γ and IFN-α/β-mediated signal	MCA-induced sarcomas Wider spectrum in p53$^{-/-}$ background[53] Mammary carcinomas[33]

6. REFERENCES

1. M.J. Smyth, K.Y. Thia, S.E. Street, E. Cretney, J.A. Trapani, M. Taniguchi, T. Kawano, S.B. Pelikan, N.Y. Crowe and D.I. Godfrey. Differential tumor surveillance by natural killer (NK) and NKT cells. *J Exp Med* **191**:661–668 (2000).

2. M. Girardi, D.E. Oppenheim, C.R. Steele, J.M. Lewis, E. Glusac, R. Filler, P. Hobby, B. Sutton, R.E. Tigelaar and A.C. Hayday. Regulation of cutaneous malignancy by gammadelta T cells. *Science* **294**:605–609 (2001).

3. L. Ruggeri, M. Capanni, M.F. Martelli and A. Velardi. Cellular therapy: exploiting NK cell alloreactivity in transplantation. *Curr Opin Hematol* **8**:355–359 (2001).

4. L. Ruggeri, M. Capanni, E. Urbani, K. Perruccio, W.D. Shlomchik, A. Tosti, S. Posati, D. Rogaia, F. Frassoni, F. Aversa, M.F. Martelli and A. Velardi. Effectiveness of donor natural killer cell alloreactivity in mismatched hematopoietic transplants. *Science* **295**:2097–2100 (2002).

5. S.E. Street, Y. Hayakawa, Y. Zhan, A.M. Lew, D. MacGregor, A.M. Jamieson, A. Diefenbach, H. Yagita, D.I. Godfrey and M.J. Smyth. Innate immune surveillance of spontaneous B cell lymphomas by natural killer cells and gammadelta T cells. *J Exp Med* **199**:879–884 (2004).

6. C. Curcio, E. Di Carlo, R. Clynes, M.J. Smyth, K. Boggio, E. Quaglino, M. Spadaro, M.P. Colombo, A. Amici, P.L. Lollini, P. Musiani and G. Forni. Nonredundant roles of antibody, cytokines, and perforin in the eradication of established Her-2/neu carcinomas. *J Clin Invest* **111**:1161–1170 (2003).

7. Z. Cui and M.C. Willingham. The effect of aging on cellular immunity against cancer in SR/CR mice. *Cancer Immunol Immunother* **53**:473–478 (2004).

8. G.P. Dunn, A.T. Bruce, K.C. Sheehan, V. Shankaran, R. Uppaluri, J.D. Bui, M.S. Diamond, C.M. Koebel, C. Arthur, J.M. White and R.D. Schreiber. A critical function for type I interferons in cancer immunoediting. *Nat Immunol* **6**:722–729 (2005).

9. M.J. Smyth. Type I interferon and cancer immunoediting. *Nat Immunol* **6**:646–648 (2005).

10. M. Colonna, A. Krug and M. Cella. Interferon-producing cells: on the front line in immune responses against pathogens. *Curr Opin Immunol* **14**:373–379 (2002).

11. J.P. Houchins, T. Yabe, C. McSherry and F.H. Bach. DNA sequence analysis of NKG2, a family of related cDNA clones encoding type II integral membrane proteins on human natural killer cells. *J Exp Med* **173**:1017–1020 (1991).

12. S. Bauer, V. Groh, J. Wu, A. Steinle, J.H. Phillips, L.L. Lanier and T. Spies. Activation of NK cells and T cells by NKG2D, a receptor for stress-inducible MICA. *Science* **285**:727–729 (1999).

13. A. Cerwenka, A.B. Bakker, T. McClanahan, J. Wagner, J. Wu, J.H. Phillips and L.L. Lanier. Retinoic acid early inducible genes define a ligand family for the activating NKG2D receptor in mice. *Immunity* **12**:721–727 (2000).

14. A. Diefenbach, A.M. Jamieson, S.D. Liu, N. Shastri and D.H. Raulet. Ligands for the murine NKG2D receptor: expression by tumor cells and activation of NK cells and macrophages. *Nat Immunol* **1**:119–126 (2000).

15. A. Cerwenka and L.L. Lanier. Ligands for natural killer cell receptors: redundancy or specificity. *Immunol Rev* **181**:158–169 (2001).

16. H.A. Stephens. MICA and MICB genes: can the enigma of their polymorphism be resolved? *Trends Immunol* **22**:378–385 (2001).

17. D. Cosman, J. Mullberg, C.L. Sutherland, W. Chin, R. Armitage, W. Fanslow, M. Kubin and N.J. Chalupny. ULBPs, novel MHC class I-related molecules, bind to CMV glycoprotein UL16 and stimulate NK cytotoxicity through the NKG2D receptor. *Immunity* **14**:123–133 (2001).

18. L.N. Carayannopoulos, O.V. Naidenko, D.H. Fremont and W.M. Yokoyama. Cutting edge: murine UL16-binding protein-like transcript 1: a newly described tran-

script encoding a high-affinity ligand for murine NKG2D. *J Immunol* **169**:4079–4083 (2002).

19. V.M. Braud, D.S. Allan and A.J. McMichael. Functions of nonclassical MHC and non-MHC-encoded class I molecules. *Curr Opin Immunol* **11**:100–108 (1999).

20. V. Groh, S. Bahram, S. Bauer, A. Herman, M. Beauchamp and T. Spies. Cell stress-regulated human major histocompatibility complex class I gene expressed in gastro-intestinal epithelium. *Proc Natl Acad Sci USA* **93**:12445–12450 (1996).

21. J.A. Hamerman, K. Ogasawara and L.L. Lanier. Cutting edge: Toll-like receptor signaling in macrophages induces ligands for the NKG2D receptor. *J Immunol* **172**:2001–2005 (2004).

22. D.H. Raulet. Roles of the NKG2D immunoreceptor and its ligands. *Nat Rev Immunol* **3**:781–790 (2003).

23. A. Rolle, M. Mousavi-Jazi, M. Eriksson, J. Odeberg, C. Soderberg-Naucler, D. Cosman, K. Karre and C. Cerboni. Effects of human cytomegalovirus infection on ligands for the activating NKG2D receptor of NK cells: up-regulation of UL16-binding protein (ULBP)1 and ULBP2 is counteracted by the viral UL16 protein. *J Immunol* **171**:902–908 (2003).

24. S. Gasser, S. Orsulic, E.J. Brown and D.H. Raulet. The DNA damage pathway regulates innate immune system ligands of the NKG2D receptor. *Nature* **436**:1186–1190 (2005).

25. A.M. Jamieson, A. Diefenbach, C.W. McMahon, N. Xiong, J.R. Carlyle and D.H. Raulet. The role of the NKG2D immunoreceptor in immune cell activation and natural killing. *Immunity* **17**:19–29 (2002).

26. D. Pende, P. Rivera, S. Marcenaro, C.C. Chang, R. Biassoni, R. Conte, M. Kubin, D. Cosman, S. Ferrone, L. Moretta and A. Moretta. Major histocompatibility complex class I-related chain A and UL16-binding protein expression on tumor cell lines of different histotypes: analysis of tumor susceptibility to NKG2D-dependent natural killer cell cytotoxicity. *Cancer Res* **62**:6178–6186 (2002).

27. A. Diefenbach, E.R. Jensen, A.M. Jamieson and D.H. Raulet. Rae1 and H60 ligands of the NKG2D receptor stimulate tumour immunity. *Nature* **413**:165–171 (2001).

28. Y. Hayakawa, J.M. Kelly, J.A. Westwood, P.K. Darcy, A. Diefenbach, D. Raulet and M.J. Smyth. Cutting edge: tumor rejection mediated by NKG2D receptor–ligand interaction is dependent upon perforin. *J Immunol* **169**:5377–5381 (2002).

29. M.J. Smyth, J. Swann, J.M. Kelly, E. Cretney, W.M. Yokoyama, A. Diefenbach, T.J. Sayers and Y. Hayakawa. NKG2D recognition and perforin effector function mediate effective cytokine immunotherapy of cancer. *J Exp Med* **200**:1325–1335 (2004).

30. J.A. Westwood, J.M. Kelly, J.E. Tanner, M.H. Kershaw, M.J. Smyth and Y. Hayakawa. Cutting edge: novel priming of tumor-specific immunity by NKG2D-triggered NK cell-mediated tumor rejection and Th1-independent CD4+ T cell pathway. *J Immunol* **172**:757–761 (2004).

31. M.J. Smyth, D.I. Godfrey and J.A. Trapani. A fresh look at tumor immunosurveillance and immunotherapy. *Nat Immunol* **2**:293–299 (2001).

32. M.J. Smyth, J. Swann, E. Cretney, N. Zerafa, W.M. Yokoyama and Y. Hayakawa. NKG2D function protects the host from tumor initiation. *J Exp Med* **202**:583–588 (2005).

33. V. Shankaran, H. Ikeda, A.T. Bruce, J.M. White, P.E. Swanson, L.J. Old and R.D. Schreiber. IFNgamma and lymphocytes prevent primary tumour development and shape tumour immunogenicity. *Nature* **410**:1107–1111 (2001).

34. M.J. Smyth, N.Y. Crowe and D.I. Godfrey. NK cells and NKT cells collaborate in host protection from methylcholanthrene-induced fibrosarcoma. *Int Immunol* **13**:459–463 (2001).

35. N.Y. Crowe, M.J. Smyth and D.I. Godfrey. A critical role for natural killer T cells in immunosurveillance of methylcholanthrene-induced sarcomas. *J Exp Med* **196**:119–127 (2002).

36. N.Y. Crowe, J.M. Coquet, S.P. Berzins, K. Kyparissoudis, R. Keating, D.G. Pellicci, Y. Hayakawa, D.I. Godfrey and M.J. Smyth. Differential antitumor immunity mediated by NKT cell subsets in vivo. *J Exp Med* **202**:1279–1288 (2005).

37. V. Groh, J. Wu, C. Yee and T. Spies. Tumour-derived soluble MIC ligands impair expression of NKG2D and T-cell activation. *Nature* **419**:734–738 (2002).

38. H.R. Salih, H. Antropius, F. Gieseke, S.Z. Lutz, L. Kanz, H.G. Rammensee and A. Steinle. Functional expression and release of ligands for the activating immunoreceptor NKG2D in leukemia. *Blood* **102**:1389–1396 (2003).

39. D.E. Oppenheim, S.J. Roberts, S.L. Clarke, R. Filler, J.M. Lewis, R.E. Tigelaar, M. Girardi and A.C. Hayday. Sustained localized expression of ligand for the activating NKG2D receptor impairs natural cytotoxicity in vivo and reduces tumor immunosurveillance. *Nat Immunol* **6**:928–937 (2005).

40. A. Moretta, C. Bottino, M. Vitale, D. Pende, C. Cantoni, M.C. Mingari, R. Biassoni and L. Moretta. Activating receptors and coreceptors involved in human natural killer cell-mediated cytolysis. *Annu Rev Immunol* **19**:197–223 (2001).

41. S.A. Rosenberg. Progress in human tumour immunology and immunotherapy. *Nature* **411**:380–384 (2001).

42. M.J. Smyth, M. Taniguchi and S.E. Street. The anti-tumor activity of IL-12: mechanisms of innate immunity that are model and dose dependent. *J Immunol* **165**:2665–2670 (2000).

43. J. Brady, Y. Hayakawa, M.J. Smyth and S.L. Nutt. IL-21 induces the functional maturation of murine NK cells. *J Immunol* **172**:2048–2058 (2004).

44. H. Okamura, H. Tsutsui, S. Kashiwamura, T. Yoshimoto and K. Nakanishi. Interleukin-18: a novel cytokine that augments both innate and acquired immunity. *Adv Immunol* **70**:281–312 (1998).

45. Y.I. Son, R.M. Dallal, R.B. Mailliard, S. Egawa, Z.L. Jonak and M.T. Lotze. Interleukin-18 (IL-18) synergizes with IL-2 to enhance cytotoxicity, interferon-gamma production, and expansion of natural killer cells. *Cancer Res* **61**:884–888 (2001).

46. G. Wang, M. Tschoi, R. Spolski, Y. Lou, K. Ozaki, C. Feng, G. Kim, W.J. Leonard and P. Hwu. In vivo antitumor activity of interleukin 21 mediated by natural killer cells. *Cancer Res* **63**:9016–9022 (2003).

47. H.L. Ma, M.J. Whitters, R.F. Konz, M. Senices, D.A. Young, M.J. Grusby, M. Collins and K. Dunussi-Joannopoulos. IL-21 activates both innate and adaptive immunity to generate potent antitumor responses that require perforin but are independent of IFN-gamma. *J Immunol* **171**:608–615 (2003).

48. A. Moroz, C. Eppolito, Q. Li, J. Tao, C.H. Clegg and P.A. Shrikant. IL-21 enhances and sustains CD8+ T cell responses to achieve durable tumor immunity: comparative evaluation of IL-2, IL-15, and IL-21. *J Immunol* **173**:900–909 (2004).

49. M.T. Kasaian, M.J. Whitters, L.L. Carter, L.D. Lowe, J.M. Jussif, B. Deng, K.A. Johnson, J.S. Witek, M. Senices, R.F. Konz, A.L. Wurster, D.D. Donaldson, M. Collins, D.A. Young and M.J. Grusby. IL-21 limits NK cell responses and promotes antigen-specific T cell activation: a mediator of the transition from innate to adaptive immunity. *Immunity* **16**:559–569 (2002).

50. R. Takaki, Y. Hayakawa, A. Nelson, P.V. Sivakumar, S. Hughes, M.J. Smyth and L.L. Lanier. IL-21 enhances tumor rejection through a NKG2D-dependent mechanism. *J Immunol* **175**:2167–2173 (2005).

51. R. Zeng, R. Spolski, S.E. Finkelstein, S. Oh, P.E. Kovanen, C.S. Hinrichs, C.A. Pise-Masison, M.F. Radonovich, J.N. Brady, N.P. Restifo, J.A. Berzofsky and W.J. Leonard. Synergy of IL-21 and IL-15 in regulating CD8+ T cell expansion and function. *J Exp Med* **201**:139–148 (2005).

52. M.J. Smyth, M.E. Wallace, S.L. Nutt, H. Yagita, D.I. Godfrey and Y. Hayakawa. Sequential activation of NKT cells and NK cells provides effective innate immunotherapy of cancer. *J Exp Med* **201**:1973–1985 (2005).

53. D.H. Kaplan, V. Shankaran, A.S. Dighe, E. Stockert, M. Aguet, L.J. Old and R.D. Schreiber. Demonstration of an interferon gamma-dependent tumor surveillance system in immunocompetent mice. *Proc Natl Acad Sci USA* **95**:7556–7561 (1998).

54. S.E. Street, E. Cretney and M.J. Smyth. Perforin and interferon-gamma activities independently control tumor initiation, growth, and metastasis. *Blood* **97**:192–197 (2001).

55. S.E. Street, J.A. Trapani, D. MacGregor and M.J. Smyth. Suppression of lymphoma and epithelial malignancies effected by interferon gamma. *J Exp Med* **196**:129–134 (2002).

56. S. Mitra-Kaushik, J. Harding, J. Hess, R. Schreiber and L. Ratner. Enhanced tumorigenesis in HTLV-1 tax-transgenic mice deficient in interferon-gamma. *Blood* **104**:3305–3311 (2004).

57. I. Gresser, F. Belardelli, C. Maury, M.T. Maunoury and M.G. Tovey. Injection of mice with antibody to interferon enhances the growth of transplantable murine tumors. *J Exp Med* **158**:2095–2107 (1983).

8

REGULATION OF ADAPTIVE IMMUNITY BY CELLS OF THE INNATE IMMUNE SYSTEM: BONE MARROW NATURAL KILLER CELLS INHIBIT T CELL PROLIFERATION

Prachi P. Trivedi, Taba K. Amouzegar, Paul C. Roberts,
Norbert A. Wolf, and Robert H. Swanborg

1. INTRODUCTION

Natural killer (NK) cells represent the third largest population of lymphocytes after T and B cells and are derived from the same precursor cell, although they do not express antigen-specific receptors (Yokoyama et al., 2004). However, they can distinguish normal host cells from virus-infected or tumor cells and lyse the latter without prior immunological sensitization — hence the name "natural killer" cell (Trinchieri, 1989). It was determined that the NK cells recognize target cells because the latter are deficient in, or lack, the expression of host major histocompatibility (MHC) class I molecules (Karre et al., 1986).

The fate of the target cell is dependent upon membrane receptors on the NK cell that either initiate target cell killing in the case of virus-infected or tumor cells, or that recognize self MHC class I on normal cells and spare those cells from cytolysis. Activating receptors that initiate cytolytic function against tumor target cells have been identified. In humans, these include NKp46, NKp30, and NKp44, which are referred to as natural cytotoxicity receptors (Moretta et al., 2001). Their function can be blocked with appropriate monoclonal antibodies. They appear to function in concert with NKG2D, a C-type lectin receptor that interacts with molecules expressed in various tumors.

Prachi P. Trivedi, Harvard University, Cambridge, Massachusetts 02138, USA; Taba K. Amouzegar, Paul C. Roberts, Norbert A. Wolf, and Robert H. Swanborg, Wayne State University School of Medicine, Detroit, Michigan 48201, USA.

NK cells also express inhibitory receptors that bind to MHC class I on the surface of normal host cells and protect them from cytolysis. These include the Ly49 receptors in rodents (Yokoyama et al., 2004), and killer cell immunoglobulin-like receptors in humans (Long et al., 1997). Subsequent to binding to MHC class I, these signal through immunoreceptor tyrosine-based inhibitory motifs (ITIMs) and switch off the NK cells.

2. REGULATORY FUNCTION OF NK CELLS

More recently it has been reported that NK cells can interact with T cells, B cells, and dendritic cells (DCs) to regulate their activity. For example, NK cells interact in a contact-dependent fashion with mature DCs, resulting in activation and secretion of cytokines by both cell types (Ferlazzo et al., 2002; Gerosa et al., 2002; Piccioli et al., 2002). NK cells have also been shown to be necessary for allograft tolerance to pancreatic islet cells, apparently through a perforin-dependent pathway that may target DCs for cytolysis (Beilke et al., 2005). In other studies it was shown that human decidual NK cells constitute a unique subset that differs significantly from peripheral blood NK cells. The decidual NK cells have been postulated to play a role in maintaining maternal–fetal tolerance by downregulating immune responses to the fetus (Koopman et al., 2003). In contrast to peripheral NK cells, the decidual NK cells lack cytotoxic function because they fail to polarize their microtubule organizing centers and perforin granules to the NK–target cell synapse (Kopcow et al., 2005).

Evidence is also accumulating that NK cells may play a regulatory role in human and experimental autoimmune diseases. For example, there is an inverse relationship between NK cell function and disease status in patients with multiple sclerosis (MS); NK activity is high during remission and decreases during relapses (Kastrukoff et al., 2003; Munschauer et al., 1995; Takahashi et al., 2004). In the animal model of MS, depletion of NK cells from C57BL/6 mice with anti-NK1.1 antibody resulted in a more severe form of experimental autoimmune encephalomyelitis (EAE) (Zhang et al., 1997). Similar results were observed in Lewis rats, where NK cell depletion resulted in a more severe form and an increased incidence of EAE (Matsumoto et al., 1998).

Previous studies from our laboratory showed that DA rat bone marrow NK^+CD3^- cells inhibited proliferation MBP-reactive T cells to the self antigen, myelin basic protein (MBP) (Table 1), as well as the response of naive T cells to the mitogen Concanavalin A or to Phorbol-12-myristate-13 acetate (PMA) and ionomycin (Smeltz et al., 1999; Trivedi et al., 2005; Wolf and Swanborg, 2001).

Table 1. NK[+]CD3[−] Cells Inhibit Proliferation of
MBP-Specific T Cells to MBP

Culture[a]	cpm	S.D.
T cells + MBP	60,028	6,003
T cells + NK cells (160:1) + MBP	21,628	790
T cells + NK cells (80:1) + MBP	13,278	1,723
T cells + NK cells (20:1) + MBP	3,552	415

[a] 96-hr cultures (containing 5 μM MBP) pulsed with [³H] thymidine for the last 18 hr of culture.

3. NK CELLS INHIBIT BY A NON-CYTOTOXIC MECHANISM

Freshly isolated DA rat bone marrow NK[+]CD3[−] cells were purified by sorting using a FACSVantage cell sorter. In all experiments, purity exceeded 97%. Syngeneic T cells were activated in vitro with either Con A or PMA + ionomycin, in the presence or absence of bone marrow NK cells. The NK cells inhibited T cell proliferation in a dose-dependent fashion (Trivedi et al., 2005). In some experiments the NK cells were separated from the T cells by transmembrane inserts (0.4 μm pore size) to prevent cell-to-cell contact. When separated by a transmembrane, the NK cells were not able to inhibit T cell proliferation, indicating that cell-to-cell contact was required. Representative results are shown in Table 2. To further confirm whether soluble factors are involved, we stimulated T cells with Con A in the presence of supernatants from the NK/T cell co-cultures. The supernatants did not inhibit T cell proliferation. Thus, it is unlikely that soluble factors mediate inhibition.

Table 2. NK Cell Inhibition of T Cell Proliferation Is Contact Dependent

Culture[a]	cpm	S.D.
T cells alone (background)	145	24
T cells + PMA + ionomycin	117,702	17,701
T cells + NK cells + PMA + ionomycin	662	551
T cells + PMA + ionomycin/NK cells separated by Transwell	107,748	8,784

[a] 48-hr cultures (T:NK = 10:1) pulsed with [³H] thymidine for the last 18 hr of culture.

The bone marrow NK cells contain intracellular perforin, although they do not induce T cell apoptosis as determined by annexin V staining. Because the NK cells inhibit T cell proliferation, we postulated that they might inhibit proliferation by suppressing IL-2 production. However, when the supernatants from the T cell/NK cell cocultures were examined by ELISA, they were found to contain abundant concentrations of IL-2. This was confirmed using the IL-2-dependent CTLL-2 cell assay, which also revealed that the IL-2 was biologically active (Trivedi et al., 2005). Moreover, the T cells in co-culture with NK cells expressed activation markers, e.g., CD25 and VLA-4, and also exhibited increased cell size as determined by forward scatter in the FACScan flow cytometer (Trivedi et al., 2005).

Additional support for the conclusion that NK cells do not inhibit IL-2 receptor signaling in T cells derived from the finding that NK cells did not affect STAT5 activation. When IL-2 binds to the IL-2 receptor, the latter phosphorylates JAK1, which then phosphorylates the STAT5 transcription factor (Gaffen, 2001). Western blot analysis revealed that phosphorylated STAT5 was present in PMA + ionomycin-activated T cells in both the absence and presence of NK cells, indicating that IL-2–IL-2 receptor signaling was functional.

Furthermore, we also found that NK cell-mediated inhibition of T cell proliferation was reversible, since removal of the NK cells after 48 hr by staining with PE-labeled anti-NK1.1 monoclonal antibody (mAb) and magnetic bead separation using anti-PE microbeads (Miltenyi Biotec) restored the proliferative activity of the T cells. Together, these findings argue against a cytotoxic mechanism of NK cell-mediated inhibition of T cell proliferation. Moreover, since the NK cells suppress the proliferation of T cells activated by PMA + ionomycin, we can conclude that inhibition is independent of T cell receptor (TCR) engagement. This is because PMA and ionomycin activate T cells by activating protein kinase C and increasing intracellular calcium levels, respectively. These events are downstream of TCR engagement by antigen.

4. NK CELLS INHIBIT CELL CYCLE PROGRESSION

Using Propidium Iodide staining and flow cytometry we determined that most of the nonproliferating T cells in the T–NK co-cultures were arrested in the G0/G1 phase of the cell cycle. In contrast, almost 50% of T cells activated in the absence of NK cells were in S phase. This indicated that they were synthesizing DNA and undergoing cell division, consistent with the proliferative response that we observed in those cultures (Trivedi et al., 2005). The progression of cells from G1 to S phase is controlled by cyclins D and E, which complex with cyclin-dependent kinases (cdk) and induce phosphorylation of retinoblastoma protein, which in turn releases the transcription factor E2F. E2F is required for transcription of S phase genes (Bartek and Lukas, 2001). Using Western blot analysis and immunofluorescence staining, we determined that NK cells did not affect the upregulation of cyclin D3 or the phosphorylation of retinoblastoma protein. Thus, NK cells did not inhibit these two phases of the signaling pathway.

The activity of cyclin–cdk complexes is regulated by inhibitor proteins, including p21 (Coqueret, 2003). Using Western blot analysis, we determined that p21 was present at low levels in PMA + ionomycin-stimulated T cells, but was significantly elevated in the presence of NK cells. Moreover, when the NK cells were removed by magnetic bead separation, the concentration of p21 returned to that observed in T cells activated in the absence of NK cells (Trivedi et al., 2005). This correlated with restored proliferative responses in the NK-depleted T cell population, and strongly suggests that NK cells limit the clonal expansion of T cells by upregulation of p21. This could serve as a homeostatic mechanism to prevent unwanted activation of autoreactive T cells.

5. BONE MARROW-DERIVED NK CELLS HAVE UNIQUE FUNCTION

Since it has been reported that decidual NK cells are distinct from peripheral blood NK cells with respect to patterns of gene expression (Koopman et al., 2003) and cytotoxic activity (Kopcow et al., 2005), we compared purified DA rat bone marrow NK cells with splenic NK cells from the same animals.

We observed several major functional and phenotypic differences between freshly isolated bone marrow and splenic NK cells. Most significantly, the splenic NK cells failed to inhibit T cell proliferation (Trivedi et al., submitted manuscript). A representative experiment is shown in Table 3.

Table 3. Splenic NK Cells Do not Inhibit T Cell Proliferation

Culture[a]	cpm	S.D.
T cells alone (background)	199	25
T cells + PMA + ionomycin	71,720	440
T cells + splenic NK cells (10:1) + PMA + ionomycin	73,738	1,926
T cells + splenic NK cells (1:1) + PMA + ionomycin	69,017	2041

[a]48-hr cultures pulsed with [³H] thymidine for the last 18 hr of culture.

Phenotypically, freshly isolated splenic and bone marrow NK cells display several differences. The splenic NK cells are NK1.1[bright], CD56[bright] and do not express CD11b. They also transcribe interferon-gamma (IFN-γ) mRNA but do not secrete the protein. In contrast, bone marrow NK cells are predominantly NK1.1[dim], CD56[dim] and express CD11b. However, they do not transcribe IFN-γ messages. A small population resembles the splenic NK cells with respect to expressing the bright phenotype, but the dim bone marrow NK cells are implicated in suppression of T cell proliferation. Both splenic and bone marrow NK cells contain intracellular perforin (Trivedi et al., submitted manuscript). It will

be of interest to determine whether activated NK cells display different phenotypic or functional properties.

6. ROLE OF NK CELLS IN IMMUNE HOMEOSTASIS

It is well established that NK cells are important effector cells of the innate immune system, lysing virus-infected cells and malignant host cells that are deficient in MHC class I. However, other evidence points to a role for NK cells in
regulating adaptive immune responses. For example, NK cells have been shown
to negatively regulate T cell responses to viral infection in mice (Su et al., 2001)
and to regulate antibody production by B cells (Takeda and Dennert, 1993). NK
cells have also been shown to regulate autoimmune diseases, including uveitis in
patients (Li et al., 2005) and EAE in rodents (Matsumoto et al., 1998; Zhang et
al., 1997). Recently, it was reported that NK cell depletion enhanced EAE in
SJL/J mice, and that the NK cells were cytotoxic for myelin antigen-specific
encephalitogenic T cells in vitro using a flow cytometric assay (Xu et al., 2005).
We also saw evidence for cytotoxicity in the rat model, using the same assay
with partially enriched NK cells (Smeltz et al., 1999), but when we attempted to
repeat the same studies with highly purified, sorted NK cells, no cytotoxic activity could be detected. Thus, we concluded that cytotoxicity could have been
mediated by non-NK cells present in the enriched (but not pure) rat NK cell
preparations. Based on evidence presented above, and recently published
(Trivedi et al., 2005), we conclude the suppressive activity of NK cells in rats is
due to p21-mediated cell cycle arrest in the T cell population.

It is clear from several studies (Koopman et al., 2003; Kopcow et al., 2005;
Li et al., 2005), and this report, that functionally distinct NK cell subsets exist.
Evidence for distinct NKT cell subsets has also recently been reported (Crowe et
al., 2005). Although the biological significance of regulatory NK cells remains
to be determined, it is likely that they play a role in fetal–maternal tolerance
(Koopman et al., 2003), and in remissions from human diseases with suspected
autoimmune etiology (Kastrukoff et al., 2003; Li et al., 2005; Takahashi et al.,
2004).

However, a broader role for NK cells in immune homeostasis can be envi-
sioned. For example, it has recently been reported that bone marrow functions as
a major reservoir for memory T cells (Becker et al., 2005; Mazo, 2005). Since
the regulatory NK cells that we have found seem to be restricted to the bone
marrow subpopulation, it is conceivable that they may play a role in maintaining
bone marrow memory T cells in a quiescent state until needed for recall responses. This is currently under investigation in our laboratory.

7. ACKNOWLEDGMENTS

The work described in this article was supported by NIH grants NS06985-37 and NS048070-02.

8. REFERENCES

J. Bartek and J. Lukas. Pathways governing G1/S transition and their response to DNA damage. *FEBS Lett* **490**:117–122 (2001).

T.C. Becker, S.M. Coley, E.J. Wherry and R. Ahmed. Bone marrow is a preferred site for homeostatic proliferation of memory CD8 T cells. *J Immunol* **174**:1269–1273 (2005).

J.N. Beilke, N.R. Kuhl, Kaer Van L. and R.G. Gill. NK cells promote islet allograft tolerance via a perforin-dependent mechanism. *Nat Med* **11**:1059–1065 (2005).

O. Coqueret. New roles for p21 and p27 cell-cycle inhibitors: a function for each cell compartment? *Trends Cell Biol* **13**:65–70 (2003).

N.Y. Crowe, J.M. Coquet, S.P. Berzins, K. Kyparissoudis, R. Keating, D.G. Pellicci, Y. Hayakawa, D.I. Godfrey and M.J. Smyth. Differential antitumor immunity mediated by NKT cell subsets in vivo. *J Exp Med* **202**:1279–1288 (2005).

G. Ferlazzo, M.L. Tsang, L. Moretta, G. Melioli, R.M. Steinman and C. Munz. Human dendritic cells activate resting natural killer (NK) cells and are recognized via the NKp30 receptor by activated NK cells. *J Exp Med* **195**:343–351 (2002).

S.L. Gaffen. Signaling domains of the interleukin 2 receptor. *Cytokine* **14**:63–77 (2001).

F. Gerosa, B. Baldani-Guerra, C. Nisii, V. Marchesini, G. Carra and G. Trinchieri. Reciprocal activating interaction between natural killer cells and dendritic cells. *J Exp Med* **195**:327–333 (2002).

K. Karre, H.-G. Ljunggren, G. Piontek and R. Kiessling. Selective rejection of H-2-deficient lymphoma variants suggests alternative immune defense strategy. *Nature* **319**:675–678 (1986).

L.F. Kastrukoff, A. Lau, R. Wee, D. Zecchini, R. White and D.W. Paty. Clinical relapses of multiple sclerosis are associated with "novel" valleys in natural killer cell functional activity. *J Neuroimmunol* **145**:103–114 (2003).

L.A. Koopman, H.D. Kopcow, B. Rybalov, J.E. Boyson, J.S. Orange, F. Schatz, R. Masch, C.J. Lockwood, A.D. Schachter, P.J. Park and J.L. Strominger. Human decidual natural killer cells are a unique NK cell subset with immunomodulatory potential. *J Exp Med* **198**:1201–1212 (2003).

H.D. Kopcow, D.S.J. Allan, X. Chen, B. Rybalov, M.M. Andzeim, B. Ge and J.L. Strominger. Human decidual NK cells form immature activating synapses and are not cytotoxic. *Proc Natl Acad Sci USA* **102**:15563–15568 (2005).

Z. Li, K.L. Lim, S.P. Mahesh, B. Liu and R.B. Nussenblatt. In vivo blockade of human IL-2 receptor induces expansion of $CD56^{bright}$ regulatory NK cells in patients with active uveitis. *J Immunol* **174**:5187–5191 (2005).

E.O. Long, D.N. Burshtyn, W.P. Clark, M. Peruzzi, S. Rajagopalan, S. Rojo, N. Wagtmann and C.C. Winter. Killer cell inhibitory receptors: diversity, specificity and function. *Immunol Rev* **155**:135–144 (1997).

Y. Matsumoto, K. Kohyama, Y. Aikawa, T. Shin, Y. Kawazoe, Y. Suzuki and N. Ta-
    numa. Role of natural killer cells and TCRgd T cells in acute autoimmune encepha-
    lomyelitis. *Eur J Immunol* **28**:1681–1688 (1998).
I.B. Mazo, M. Honczarenko, H. Leung, L.L. Cavanaugh, R. Bonasio, W. Weninger, K.
    Engelke, L. Xia, R.P. EmEver, P.A. Koni, L.E. Silberstein and U.H. von Andrian.
    Bone marrow is a major reservoir and site of recruitment for central memory CD8+
    T cells. *Immunity* **22**:259–270 (2005).
A. Moretta, C. Bottino, M. Vitale, D. Pende, C. Cantoni, M.C. Mingari, R. Biassoni and
    L. Moretta. Activating receptors and coreceptors involved in human natural killer
    cell-mediated cytolysis. *Annu Rev Immunol* **11**:197–223 (2001).
F.E. Munschauer, L.A. Hartrich, C.C. Stewart and L. Jacobs. Circulating natural killer
    cells but not cytotoxic T lymphocytes are reduced in patients with active relapsing
    multiple sclerosis and little clinical disability as compared to controls. *J Neuroim-
    munol* **62**:177–181 (1995).
D. Piccioli, S. Sbrana, E. Melandri and N.M. Valiante. Contact-dependent stimulation
    and inhibition of dendritic cells by natural killer cells. *J Exp Med* **195**:335–341
    (2002).
R.B. Smeltz, N.A. Wolf and R.H. Swanborg. Inhibition of autoimmune T cell responses
    in the DA rat by bone marrow-derived natural killer cells in vitro: implications for
    autoimmunity. *J Immunol* **163**:1390–1398 (1999).
H.C. Su, K.B. Nguyen, T.P. Salazar-Mather, M.C. Ruzek, M.Y. Dalod and C.A. Biron.
    NK cell functions restrain T cell responses during viral infections. *Eur J Immunol*
    31:3048–3055 (2001).
K. Takahashi, T. Aranami, M. Endoh, S. Miyake and T. Yamamura. The regulatory role
    of natural killer cells in multiple sclerosis. *Brain* **127**:1917–1927 (2004).
K. Takeda and G. Dennert. The development of autoimmunity in C57BL/6 *lpr* mice cor-
    relates with the disappearance of natural killer type 1-positive cells: evidence for
    their suppressive action on bone marrow stem cell proliferation, B cell immu-
    noglobulin secretion, and autoimmune symptoms. *J Exp Med* **177**:155–164 (1993).
G. Trinchieri. Biology of natural killer cells. *Adv Immunol* **47**:187–376 (1989).
P.P. Trivedi, P.C. Roberts, N.A. Wolf and R.H. Swanborg. NK cells inhibit T cell prolif-
    eration via p21-mediated cell cycle arrest. *J Immunol* **174**:4590–4597 (2005).
N.A. Wolf and R.H. Swanborg. DA rat NK+CD3- cells inhibit autoreactive T cell re-
    sponses. *J Neuroimmunol* **119**:81–87 (2001).
W. Xu, G. Fazekas, H. Hara and T. Tabira. Mechanism of natural killer cell regulatory
    role in experimental autoimmune encephalomyelitis. *J Neuroimmunol* **163**:24–30
    (2005).
W.M. Yokoyama, S. Kim and A.R. French. The dynamic life of natural killer cells. *Annu
    Rev Immunol* **22**:405–429 (2004).
B.-N. Zhang, T. Yamamura, T. Kondo, M. Fujiwara and T. Tabira. Regulation of ex-
    perimental autoimmune encephalomyelitis by natural killer (NK) cells. *J Exp Med*
    186: 1677–1687 (1997).

9

INDUCTION AND MAINTENANCE OF CD8+ T CELLS SPECIFIC FOR PERSISTENT VIRUSES

Ester M.M. van Leeuwen, Ineke J.M. ten Berge,
and René A.W. van Lier

The development of immunological memory is a unique property of the adaptive immune system. Until recently most studies on the induction of virus-specific memory CD8[+] T cells have been performed in mice models of acute infection. Based on these studies, certain properties have been attributed to memory CD8[+] T cells concerning their responsiveness to antigenic stimulation and their ability to survive. However, many relevant human viruses are persistent and reach a latency stage in which there is equilibrium between the virus and the host immune system.

Analysis of virus-specific CD8[+] T cells responding to persistent viruses has now shown that these cells contrast in many aspects to the classical memory cells specific for cleared viruses, which may be explained by the different tasks that have to be fulfilled. Whereas memory CD8[+] T cells specific for cleared viruses only need to become activated upon reinfection with the same virus, CD8[+] T cells specific for persistent viruses continuously have to keep the virus under control in the latency stage. In addition, they have to adapt to possible changes in the immunocompetence of the host (e.g., during aging or at the start of immunosuppressive therapy).

We here review the recently described differences in induction and maintenance of memory CD8[+] T cells recognizing cleared versus persistent viruses both in human and in mice. We also discuss whether CD8[+] T cells specific for persistent viruses can actually be categorized as memory cells or should be considered "vigilant resting effector cells."

Ester M.M. van Leeuwen and Rene A.W. van Lier, Dept. of Experimental Immunology, Academic Medical Center, Meibergdreef 9, 1105 AZ Amsterdam, The Netherlands. R.vanlier@amc.uva.nl
Ineke J.M. ten Berge, Department of Internal Medicine, Academic Medical Center, Amsterdam.

1. PERSISTENT VIRUSES ARE PREVALENT IN HUMAN AND MICE

Persistent viruses are not cleared by the immune system, like after infection with an acute virus, but instead a balance is formed between the host immune system and the virus. Hereafter we will use the terms acute or cleared versus persistent, chronic, or latent to refer to viruses that are eliminated by the immune system or persist in the host respectively. Although many cells play a role in keeping persistent viruses under control, in this review we will focus on the role of $CD8^+$ memory T cells. Whereas memory $CD8^+$ T cells that arise after recovery of infection with an acute virus have to protect the host from secondary infections with the same pathogen, memory $CD8^+$ T cells specific for persistent viruses continuously have to prevent reactivation of the latent virus. In humans, such long-term control by $CD8^+$ T cells has been studied after infection with cytomegalovirus (CMV), Epstein-Barr virus (EBV), hepatitis B virus (HBV), hepatitis C virus (HCV), and human immunodeficiency virus (HIV). In healthy individuals, CMV and EBV are apparently well controlled, as appears from the typical absence of signs of infection. This changes when the immune system becomes compromised, i.e., after bone marrow transplantation, solid organ transplantation, or in HIV patients. Then the equilibrium between the virus and the host immune system is disturbed and the virus can easily reactivate and cause serious secondary illness.

Persistent viruses like those described above differ in the type of persistence they establish in the host. EBV is a virus that becomes latent and reactivates only periodically, resulting in limited T cell stimulation. Other viruses, like CMV, cause a more smoldering infection with ongoing low-level viral replication. T cells specific for CMV will thus frequently be activated by antigen. Infection with HIV, HCV, or HBV leads to viremia and persisting high levels of viral replication resulting in continuous T cell stimulation. Acute infections with HIV, HBV, and HCV are more frequently associated with clinical problems. This might be caused by the high ability of those viruses to mutate and escape T cell control, whereas EBV and CMV are genetically stable viruses.

Mice models are useful to study viral infections because they allow a more extensive analysis of immune responses than is possible in man. Especially, analysis of T cells in humans is mostly restricted to the peripheral blood compartment, whereas in mice cells can be retrieved from different organs. Concerning persistent viruses, murine γ-herpesvirus 68 (γHV68) infection is applied as a model for EBV infection. CMV is species specific, but still valuable information can be obtained from studies with murine CMV (MCMV). Murine polyoma virus induces a similar low-level persistent viral infection. Lymphocytic choriomeningitis virus (LCMV) is an interesting virus because different strains are either cleared (LCMV Armstrong or LCMV-WE) or cause persistent infection (LCMV clone 13, LCMV-T1b or LCMV-Docile), allowing direct comparison of T cell responses to these different types of viruses.

Thus a wide variety of persisting viruses exists, both in human and in mice. These viruses differ in tropism, behavior during persistence, immunomodulatory capacity, and genetic stability. Still, as will become apparent later, the immune reaction to different persisting viruses shows a considerable degree of similarity.

2. GENERAL EFFECTS OF PERSISTENT VIRUSES ON THE HOST IMMUNE SYSTEM

Because of their persisting nature, control of latent viruses demands much attention of the immune system. This is very evident in the case of the human herpes viruses, which will reactivate and eventually cause disease, as soon as for whatever reason T cell immunity wanes. As a consequence, infection with a persistent virus has a major impact on the composition of the T cell pool. Especially infection with CMV exerts a profound effect on both the total CD8$^+$ and CD4$^+$ T cell populations. Khan et al. described that CMV latency led to a greater clonality of the T cell repertoire[1]. Several other studies showed that mainly an expansion in far differentiated, resting cytotoxic T cells was induced in CMV carriers[2-4]. A significant increase in CD8$^+$ effector type T cells was only dependent on infection with CMV and was not related to several other pathogens studied, not even to HIV[5,6]. Within the CD4$^+$ T cell compartment, a population of cytotoxic CD28$^-$ T cells was present only in CMV-infected individuals and indeed partially responded to CMV-specific stimulation by either proliferation or cytokine production[7]. Whether all these cells are directly responsive to CMV itself or are somehow indirectly induced by CMV infection remains to be established.

CMV seems to be unique in its capacity to induce such prominent changes in the immune system, although a recent paper showed that also HCV infection had some influence on the CD8$^+$ T cell population[8]. The reason that CMV infection in particular induces these adaptations of the immune system is largely unknown. A factor that could be involved is that CMV is a virus with strong immune-evasive properties and during evolution there have been mutual adaptations between the virus and the human immune system[9,10]. Therefore it can be imagined that the immune system has to put more effort into control of CMV compared to other viruses. Another reason could be the quite large reservoir of CMV during the latency stage, as CMV establishes latency in cells of the myeloid lineage and in endothelial cells that are both abundantly present. Consequently, T cells will regularly come into contact with CMV antigens, not only because the virus undergoes continuous low-level replication but also because of its abundant spread within the body.

An additional specific feature of CMV is that it impairs the response to other infections. In old age the numbers of EBV-specific cells have been shown to increase, but this does not occur in CMV-seropositive donors[11]. We have recently shown that in CMV-seronegative renal transplant recipients the percentages of EBV-, FLU-, and RSV-specific CD8⁺ T cells increased in time. In contrast, this rise did not occur in CMV-seropositive renal transplant patients (van Leeuwen et al., submitted). In our study, CMV-specific cells increased in percentage after transplantation, which correlates with the finding that also murine CMV-specific CD8⁺ T cells continuously accumulate over time[12]. Data from the group of Welsh showed that a new infection caused permanent loss of pre-existing memory cells due to competition[13]. In view of the aforementioned data, it is remarkable that memory T cell loss was more profound during persistent infection than after infection with a cleared virus, and was a continuously ongoing process[14]. In humans with a primary CMV infection we similarly found that EBV-specific cells decreased in percentage and absolute numbers as soon as CMV-specific cells appeared in the circulation. Once the viral load became undetectable and the numbers of CMV-specific cells declined, the percentage of EBV-specific cells remained low, although the absolute number did return to the original value (van Leeuwen et al., submitted). The finding that the number of EBV-specific T cells is compromised by CMV infection is interesting in light of the observation that post-transplant lymphoproliferative disease (PTLD), an EBV-associated disorder in transplant patients, occurs more often in patients undergoing a primary CMV infection[15]. Altogether these observations imply that competition between memory cells, either for space or homeostatic cytokines, is taking place. Apparently, the homeostasis of memory T cells is differentially regulated when cells with different specificities enroll in the memory T cell pool. CD8⁺ T cells specific for persistent viruses seem to have an advantage in that situation, possibly because they are frequently triggered by their specific antigen and thereby maintained in higher numbers.

3. GENERATION OF CD8+ MEMORY T CELLS

The development of memory CD8⁺ T cells in response to infection with acute viruses has extensively been studied in mice (reviewed in [16]). Upon activation by an antigen-presenting cell bearing the antigenic peptide, naive T cells differentiate and expand until a large pool of effector cells is formed. These effector cells can eliminate virally infected cells by means of their cytotoxic capacity. Once the virus is cleared, most effector cells die and only 5–10% survive and develop into long-lived memory cells. Upon a secondary encounter with a virus the immune system will therefore be able to respond faster and more efficiently, thereby limiting cytopathic viral effects. Recently, expression of the IL-7Rα (CD127) was found to identify the memory precursor cells that survived the contraction phase and all memory cells homogeneously expressed this receptor[17,18]. Several studies have suggested that the expansion and differentiation program of CD8⁺ T cells specific for cleared viruses is already imprinted shortly after antigenic stimulation[19-21]. Moreover, in a number of models it has been shown that CD4⁺ T cell help is required for the development and maintenance of a functional CD8[+] memory T cell population that can mount a proper secondary response[22-25].

Memory T cell generation in persistent viral infections has not widely been studied. The initial response to a persistent virus is similar to that of an acute virus, with the induction of a large pool of antigen-specific effector cells, which in time contracts to leave a smaller memory population. However, CD8[+] T cells specific for persistent viruses are in many aspects very different in the memory phase when the virus persists at a lower level. After clearance of a virus, the epitope hierarchy found during acute viral infection is conserved in the memory phase[26]. In chronic infections, however, several studies showed a shift in the immunodominance of CD8[+] T cell epitopes in the memory phase as compared to the acute phase. Epitopes that were subdominant during the acute response were dominant during the memory phase and vice versa, completely changing the hierarchy of T cell epitopes. This has been described for chronic LCMV infection[27-29], and although many factors are involved in determining immunodominancy[30], it was shown that increasing the height of the viral load of a LCMV strain that is cleared from the host also induced significant changes in the epitope hierarchy of LCMV-specific CD8[+] memory T cells[29]. Comparable skewing of the immunodominant epitopes was also described in persistent murine γ-herpesvirus infection[31], for persistent mouse hepatitis virus[32], and for mouse polyoma virus[33]. Also in MCMV infection the ratio between CD8[+] T cells with different specificities changed, mainly because a few accumulated over time whereas others did not[12].

Differences in frequencies of CD8[+] T cells to dominant and subdominant epitopes were also found in human EBV infection, where the memory population was not a scaled-down version of the primary response either[34]. Other longitudinal studies in humans on the frequencies of virus-specific CD8[+] T cells recognizing different epitopes have not been performed yet. This can be explained by the difficulty to trace primary viral infections in humans and to the limitations in the available tetramers, which makes it hard to analyze CD8[+] T cells with multiple specificities in the same donor. Another complicating factor occurs in HIV infection, where the high incidence of mutations in the virus affects the dominance in the T cell response[35].

For an understanding of the development of memory T cells in persistent viral infections, it is important to know why T cells with other specificities are more prominent in the persistence phase than early after infection. An obvious explanation would be that during the latency phase the viral antigens, and thus the epitopes that are presented to the T cells, are different from those in the acute phase. Certainly, both the virus and the T cells are faced with an altered situation compared to the acute response. The virus is not constantly replicating and infecting cells anymore, but only tries to maintain latency, whereas the T cells adapt to their surveillance task. Also, these changes could account for the alterations in the composition of the oligoclonal T cell population in the phase of viral persistence.

4. FUNCTION OF MEMORY CD8+ T CELLS SPECIFIC
FOR PERSISTENT VIRUSES

Given that most studies on the functionality of memory CD8[+] T cells were performed after clearance of an acute virus, the properties that were attributed to memory cells are also derived from these studies. According to these criteria, memory CD8[+] T cells should be able to rapidly proliferate upon antigenic-stimulation and quickly produce cytokines like IL-2 and IFN-γ, chemokines like RANTES, and cytotoxic effector molecules like perforin and granzyme B[26,36–38]. Consequently, in the memory phase after the virus has been cleared there is excellent concordance between the number of antigen-specific cells measured by MHC tetramers and the number measured by intracellular cytokine staining[26]. Although memory T cells showed poor in-vitro killing as measured by ^{51}Cr-release assays[36], in vivo they were able to directly eliminate target cells only slightly slower than effector cells during the acute response[39].

Data from murine studies indicated that chronic viral infection could induce functional impairment (exhaustion) or even physical deletion of memory CD8[+] T cells specific for persistent viruses (reviewed in [40]). Direct comparison of memory CD8[+] T cells specific for acute or persistent LCMV strains indeed revealed that in chronically infected mice memory T cells lost functionality. Concerning cytokine secretion, memory cells specific for persistent viruses produced less IFN-γ, TNF-γ and IL-2[28,29,41,42]. Also, the ability to lyse target cells in vitro was less in memory cells during persistent LCMV infection[28], although efficient in-vivo killing is reported for polyoma-specific memory T cells[43]. Interestingly, the functional impairment of LCMV-specific memory CD8[+] T cells occurred in a hierarchical fashion during chronic infection. The production of IL-2 and the cytotoxic capacity were first compromised followed by the ability to secrete TNF-α. Production of IFN-γ was most resistant to functional exhaustion[28,42]. The amount of antigen present seemed to be the reason for this loss of function since a strong correlation was found between viral load and level of exhaustion. This was in line with the finding that viruses expressing excessive epitope levels induced reduced numbers of IFN-γ-producing memory cells compared to viruses with lower epitope expression[44]. Another hallmark of memory T cells is fast division upon secondary antigenic stimulation. Remarkably, in addition, in memory cells derived from persistently infected mice proliferation upon in-vitro stimulation with peptide elicited a reduced response compared to memory cells from mice infected with an acute, cleared virus[41].

From murine studies it appeared that the exhaustion of the memory T cells was directly correlated with the height of the viral load. Likewise, in humans, functional exhaustion of memory CD8[+] T cells has been described in infection with HIV and HCV, which show high viral replication and consequently high viral loads in peripheral blood, even in the chronic phase. In HIV infection, CD8[+] T cells were impaired in function[45–47] and dysfunction of HCV-specific memory cells was also reported[48,49]. The same hierarchical order in the functional impairment found in murine memory T cells seems to exist in human memory T cells since analysis of HIV-specific cells in a chronic phase of infection demonstrated that cytotoxicity and TNF-α production were impaired whereas IFN-γ could still be produced[50]. Noteworthily, for HIV infection it is still discussed whether the functional defects of HIV-specific CD8⁺ T cells are the cause or the consequence of ineffective viral control[51].

The amount of antigen is much lower in EBV and CMV infection, where during latency in healthy individuals the viral load is undetectable in peripheral blood. In these situations, the memory T cells found are still active and do not show exhaustion. Direct ex-vivo cytotoxicity has been shown for CMV-specific memory CD8⁺ T cells[52,53]. Only in elderly individuals was decreased production of IFN-γ by EBV- and CMV-specific cells found[11]. With regard to proliferation, in humans it is not possible to directly compare T cells specific for the same virus in acute and persistent form as with the LCMV strains. However, in contrast to laboratory mice, humans usually have a history of multiple infections, allowing a comparison between T cells specific for persistent viruses (e.g., CMV) and cells responding to viruses that have been cleared (e.g., influenza (FLU)). This analysis showed that proliferation upon stimulation with a specific peptide was similar for CMV-specific cells and FLU-specific cells from the same donor[54].

Thus, persistence of antigen does not always result in exhaustion of memory T cells — only when viral loads are continuously high. Therefore, it would be erroneous to describe memory cells specific for persistent viruses as functionally incompetent in general. Especially since, for example, in CMV and EBV infection, memory CD8⁺ T cells function properly in controlling the latent virus.

5. PHENOTYPE OF MEMORY CD8+ T CELLS SPECIFIC FOR PERSISTENT VIRUSES

An apparent difference in phenotype between memory cells specific for cleared viruses and viruses that are maintained in the body is found in mouse models. Here, expression of the lymph node homing receptors CCR7 and CD62L and, more recently, IL-7Rα are most frequently used to define different subsets of memory T cells. Without elaborating too much on the differences and the interrelationship between CCR7⁻CD62L⁻ cells (effector memory) and CCR7⁺CD62L⁺ cells (central memory)[16,55], one can state that the latter represent the memory cells that differentiate after clearance of a virus[56]. Although during acute infection the large majority of CD8⁺ T cells do not express IL-7Rα, all memory cells after acute infection express this receptor, and therefore it is used as a marker for memory cells[17,18].

Memory CD8⁺ T cells in persistent viral infection do have a clearly different phenotype. In chronic LCMV infection, only a few memory CD8⁺ T cells re-expressed CD62L and CCR7[29,41]. A recent report, using infection with an LCMV-strain that is cleared, described that, whereas at high naive T cell precursor frequencies a population of CCR7⁻CD62L⁻ memory CD8⁺ T cells reverted to CCR7⁺CD62L⁺, this was not the case when started with low precursor frequencies[57]. Considering that with low precursor frequencies there would be more antigen available per T cell, this would resemble a persistent infection as also there the chances of antigen encounter are high and more memory cells remain deprived of CCR7 and CD62L re-expression. In addition to the low expression of CCR7 and CD62L, expression of the memory cell marker IL-7Rα was much lower in persistently infected mice, as compared to the expression on all memory cells after clearance of an acute virus[41,58,59].

Most studies on memory T cells in humans have been performed in individuals without acute infection who have various memory cells for both eliminated and latent viruses (reviewed in [60]) . CD8⁺ T cells for human viruses have been phenotypically characterized using different combinations of cell surface molecules. CD8⁺ T cells for viruses that are cleared by the host such as influenza (FLU) and respiratory syncytial virus (RSV) uniformly express both costimulatory molecules CD28 and CD27, differ in expression of the chemokine receptor CCR7, and are CD45RA⁻. As described in mice, human memory CD8⁺ T cells specific for acute viruses all express IL-7Rα[54,61,62].

Analysis of human CD8⁺ T cells specific for persistent viruses (reviewed in [63]) revealed that these cells hardly ever express CCR7; only HCV-specific cells can be CCR7⁺. Concerning the other surface molecules, there is a large heterogeneity between cells with different viral specificities. HCV-specific cells usually express both CD28 and CD27, are CD45RA⁻, and are accordingly classified as early memory cells. CD8⁺ T cells recognizing EBV can be found as early or intermediate memory cells because the cells lack CD45RA, do express CD27, but differ in CD28 expression. Also, CD8⁺ T cells specific for HIV are mostly intermediate memory cells, usually CD28⁻CD27⁺. The virus-specific cells that are most differentiated are CMV-specific cells from which the majority can be found as late memory cells, CCR7⁻CD28⁻CD27⁻CD45RA⁺, although CMV-specific cells can also be found in other differentiation stages[8,64–72]. Interestingly, when FLU- and RSV-specific cells derived from the human lung were investigated, these showed a decreased expression of CD28 and CD27 compared to cells with the same specificity in peripheral blood[73]. This suggests that more frequent encounter with their antigen at the site of infection, like the lung for respiratory viruses, changes the phenotype of CD8⁺ T cells specific for acute viruses such that they more resemble CD8⁺ T cells specific for persistent viruses.

As was shown in mice studies, also for human memory cells it seems that the height of the viral load determines the phenotype of the memory CD8⁺ T cells. A higher number of CMV-specific cells, which likely is the result of a higher antigenic load, correlated significantly with a more CD27⁻ phenotype of these cells[74].

Expression of IL-7Rα has been analyzed on CMV-, EBV-, and HIV-specific cells, and in contrast to memory cells specific for eliminated viruses, only part of the memory cells for persistent viruses expressed this receptor long after primary infection[54,75]. This implies that, as long as antigen is present, IL-7Rα will not be re-expressed on memory T cells. Another explanation could be that T cells specific for persistent viruses have received such prolonged TCR stimulation that the downregulation of IL-7Rα has become an irreversible differentiation event. This is in line with the finding that in renal transplant recipients the maximum CMV viral load measured by PCR early after transplantation inversely correlated with the percentage of IL-7Rα[+] CMV-specific cells at a late time point[54].

6. MAINTENANCE OF MEMORY CD8+ T CELLS SPECIFIC FOR PERSISTENT VIRUSES

An important characteristic of memory CD8[+] T cells that remain after elimination of a virus is that they survive for prolonged periods devoid of contact with antigen[76,77]. To maintain a stable pool of memory T cells, extrinsic factors are required, however, for survival and homeostatic proliferation. The cytokines IL-7 and IL-15, both members of the common γ-chain family of cytokines, have in this regard been identified as essential factors[78-81]. The requirement for IL-7 is stressed by the finding that expression of IL-7Rα is selectively found on memory precursor cells, and all memory cells after elimination of an acute virus highly express this receptor[17,18].

The requirements for survival of memory CD8[+] T cells specific for persistent viruses appear to be different. First, the response to the homeostatic cytokines IL-7 and IL-15 is clearly less for memory cells specific for persistent viruses[41,54]. This can be explained by the fact that only part of the memory cells specific for persistent viruses express the IL-7Rα and expression of the IL-15Rα is also lower than on memory cells specific for cleared viruses[41,54]. Also, experiments in IL-15[-/-] mice showed that memory CD8[+] T cells specific for persisting murine γHV68 did not depend on IL-15 for their maintenance[82]. Second, after adoptive transfer of memory CD8[+] T cells specific for persistent viruses into naive mice, these memory cells do not survive and/or proliferate in the absence of their specific antigen and, therefore, were called antigen-addicted[41,82]. The same was found in murine models where treatment of chronically infected mice resulted not only in clearance of the pathogen but simultaneously in a decline in antigen-specific T cells and loss of protective immunity[83,84]. In a normal situation, however, T cells specific for persistent viruses will regularly encounter their antigen. Yet, the conclusion will remain that memory cells specific for persistent viruses probably depend on contact with antigen for their maintenance.

An attractive consideration is that antigen-specific memory CD8[+] T cells might also still develop during the persistence phase instead of only after the acute phase of infection. Support comes from experiments in which priming of epitope-specific CD8[+] T cells in bone marrow chimeric mice could be visualized during persistent viral infection[33]. This implies that not all memory cells found late in infection have been maintained for a long period but some might have been induced somewhere along the period of persistent infection. Interestingly, cells primed during the persistent phase of infection had higher CD62L expression, which could be explained by a shorter contact with most likely lower levels of antigen.

Another probably important factor in the maintenance of memory CD8[+] T cells in persistent infections is CD4[+] T cell help[33]. Their presence is known to be required for induction and maintenance of fully functional memory CD8[+] T cells[22-25] in models of infection with acute viruses. Their role in persistent infection was demonstrated in CD4[-/-] mice infected with chronic LCMV, where absolutely no IL-7Rα[+] CD8[+] memory cells could be detected[58]. In humans, the emergence of peripheral blood CMV-specific CD4[+] T cells in symptomatic primary CMV infections was delayed compared to asymptomatic infection, emphasizing their contribution to resolution of the acute phase of the infection[52]. Also in vitro, human CMV-specific cells dramatically increased their proliferation when helper cell-derived cytokines or activated CD4[+] T cells themselves were added to the culture[53,71].

Thus, presumably, memory CD8[+] T cells specific for persistent viruses require contact with antigen and CD4[+] T cell help for their maintenance.

7. REGULATION OF IL-7Rα EXPRESSION BY THE PRESENCE OF ANTIGEN

The difference in expression of IL-7Rα between memory CD8[+] T cells after clearance of an acute virus or during persistent infection and the consequences for the importance of IL-7 for the survival of these cells can be explained by the notion that IL-7Rα is downregulated upon TCR activation of the cell[54,85,86]. After clearance of an acute virus, the memory T cells are not triggered by antigen anymore and re-express the IL-7Rα, whereas in persistent infection the T cells continue to regularly encounter antigen. Indeed, during a recall response after viral clearance, the levels of IL-7Rα on the memory T cells decreased again, showing that changes in expression of IL-7Rα reflect the activation state of memory T cells[58]. That the ongoing TCR stimulation was responsible for the continued downregulation of IL-7Rα could be confirmed by treating mice with peptides that were either short lived or persisted in vivo[59].

Also in humans, the influence of the level of antigen was shown as the percentage of IL-7Rα[+] CMV-specific cells declined when the viral load increased during CMV reactivation and a higher maximum viral load inversely correlated with the percentage of IL-7Rα[+] CMV-specific cells[54]. In HIV patients, where high viral loads both for HIV and for other viruses are more common, the level of IL-7Rα expression on total CD8⁺ T cells is lower than in healthy individuals. In line with a role for contact with antigen, the percentage of IL-7Rα⁺ CD8⁺ T cells reached normal levels when the antigenic load decreased after successful treatment[75,87-89]. These data lead to the assumption that the presence of antigen downregulates the receptor and thus decreases the dependency of memory CD8⁺ T cells on IL-7.

8. CONCLUDING REMARKS

Combined, these data give rise to the hypothesis that for the maintenance of CD8⁺ T cell memory in persistent viral infections cells need regular contact with their specific antigen in combination with help signals from CD4⁺ T cells. In contrast to memory cells developed after clearance of an acute virus, there appears to be a minor role for IL-7, and likely also IL-15, in the maintenance of T cells specific for persistent viruses.

Furthermore, whether CD8⁺ T cells found during persistent infection can actually be named memory T cells can be questioned. They do not have to "memorize" the specific antigen because the virus is still present in the host and they continuously have to keep the infection under control. This could also explain why these cells, concerning function, phenotype, and maintenance requirements, differ so much from the classical memory cells that develop after acute infection when the virus is cleared. Because CD8⁺ T cells during persistent infection cannot retire as after infection with a cleared virus, but have to continue to monitor the virus, they appear to function as "vigilant resting effector cells."

9. REFERENCES

1. N. Khan, N. Shariff, M. Cobbold, R. Bruton, J.A. Ainsworth, A.J. Sinclair, L. Nayak and P.A. Moss. Cytomegalovirus seropositivity drives the CD8 T cell repertoire toward greater clonality in healthy elderly individuals. *J Immunol* **169**:1984–1992 (2002.

2. J.W. Gratama, A.M. Naipal, M.A. Oosterveer, T. Stijnen, H.C. Kluin-Nelemans, L.A. Ginsel, G.J. Ottolander, A.C. Hekker, J. D'Amaro and M van der Giessen. Effects of herpes virus carrier status on peripheral T lymphocyte subsets. *Blood* **70**:516–523 (1987).

3. E.C. Wang, P.A. Moss, P. Frodsham, P.J. Lehner, J.I. Bell and L.K. Borysiewicz. CD8highCD57+ T lymphocytes in normal, healthy individuals are oligoclonal and respond to human cytomegalovirus. *J Immunol* **155**:5046–5056 (1995).

4. T.G. Evans, E.G. Kallas, A.E. Luque, M. Menegus, C. McNair and R.J. Looney. Expansion of the CD57 subset of CD8 T cells in HIV-1 infection is related to CMV serostatus. *AIDS* **13**:1139–1141 (1999).

5. T.W. Kuijpers, M.T. Vossen, M.R. Gent, J.C. Davin, M.T. Roos, P.M. Wertheim-van Dillen, J.F. Weel, P.A. Baars and RA van Lier. Frequencies of circulating cyto-

lytic, CD45RA+CD27–, CD8+ T lymphocytes depend on infection with CMV. *J Immunol* **170**:4342–4348 (2003).

6.   V. Bekker, C. Bronke, H.J. Scherpbier, J.F. Weel, S. Jurriaans, P.M. Wertheim-van Dillen, V. van Leth, J.M. Lange, K. Tesselaar, D. van Baarle and T.W. Kuijpers. Cytomegalovirus rather than HIV triggers the outgrowth of effector CD8+CD45RA+. *AIDS* **19**:1025–1034 (2005).

7.   E.M. van Leeuwen, E.B. Remmerswaal, M.T. Vossen, A.T. Rowshani, P.M. Wertheim-van Dillen, R.A. van Lier and I.J. ten Berge. Emergence of a CD4+CD28– granzyme B+, cytomegalovirus-specific T cell subset after recovery of primary cytomegalovirus infection. *J Immunol* **173**:1834–1841 (2004).

8.   M. Lucas, A.L. Vargas-Cuero, G.M. Lauer, E. Barnes, C.B. Willberg, N. Semmo, B.D. Walker, R. Phillips and P. Klenerman. Pervasive influence of hepatitis C virus on the phenotype of antiviral CD8+ T cells. *J Immunol* **172**:1744–1753 (2004).

9.   M.J. Reddehase. Antigens and immunoevasins: opponents in cytomegalovirus immune surveillance. *Nat Rev Immunol* **2**:831–844 (2002).

10.   E.S. Mocarski Jr. Immunomodulation by cytomegaloviruses: manipulative strategies beyond evasion. *Trends Microbiol* **10**:332–339 (2002).

11.   N. Khan, A. Hislop, N. Gudgeon, M. Cobbold, R. Khanna, L. Nayak, A.B. Rickinson and P.A. Moss. Herpesvirus-specific CD8 T cell immunity in old age: cytomegalovirus impairs the response to a coresident EBV infection. *J Immunol* **173**:7481–7489 (2004).

12.   U. Karrer, S. Sierro, M. Wagner, A. Oxenius, H. Hengel, U.H. Koszinowski, R.E. Phillips and P. Klenerman. Memory inflation: continuous accumulation of antiviral CD8+ T cells over time. *J Immunol* **170**:2022–2029 (2003).

13.   R.M. Welsh, K. Bahl and X.Z. Wang. Apoptosis and loss of virus-specific CD8+ T-cell memory. *Curr Opin Immunol* **16**:271–276 (2004).

14.   S.K. Kim and R.M. Welsh. Comprehensive early and lasting loss of memory CD8 T cells and functional memory during acute and persistent viral infections. *J Immunol* **172**:3139–3150 (2004).

15.   R.C. Walker, W.F. Marshall, J.G. Strickler, R.H. Wiesner, J.A. Velosa, T.M. Habermann, C.G. McGregor and C.V. Paya. Pretransplantation assessment of the risk of lymphoproliferative disorder. *Clin Infect Dis* **20**:1346–1353 (1995).

16.   S.M. Kaech, E.J. Wherry and R. Ahmed. Effector and memory T-cell differentiation: implications for vaccine development. *Nat Rev Immunol* **2**:251–262 (2002).

17.   S.M. Kaech, J.T. Tan, E.J. Wherry, B.T. Konieczny, C.D. Surh and R. Ahmed. Selective expression of the interleukin 7 receptor identifies effector CD8 T cells that give rise to long-lived memory cells. *Nat Immunol* **4**:1191–1198 (2003).

18.   K.M. Huster, V. Busch, M. Schiemann, K. Linkemann, K.M. Kerksiek, H. Wagner and D.H. Busch. Selective expression of IL-7 receptor on memory T cells identifies early CD40L-dependent generation of distinct CD8+ memory T cell subsets. *Proc Natl Acad Sci USA* **101**:5610–5615 (2004).

19.   S.M. Kaech and R. Ahmed. Memory CD8+ T cell differentiation: initial antigen encounter triggers a developmental program in naive cells. *Nat Immunol* **2**:415–422 (2001).

20.   M.J. van Stipdonk, E.E. Lemmens and S.P. Schoenberger. Naive CTLs require a single brief period of antigenic stimulation for clonal expansion and differentiation. *Nat Immunol* **2**:423–429 (2001).

21. V.P. Badovinac, B.B. Porter and J.T. Harty. Programmed contraction of CD8(+) T cells after infection. *Nat Immunol* **3**:619–626 (2002).

22. E.M. Janssen, E.E. Lemmens, T. Wolfe, U. Christen, M.G. von Herrath and S.P. Schoenberger. CD4+ T cells are required for secondary expansion and memory in CD8+ T lymphocytes. *Nature* **421**:852–856 (2003).

23. D.J. Shedlock and H. Shen. Requirement for CD4 T cell help in generating functional CD8 T cell memory. *Science* **300**:337–339 (2003).

24. J.C. Sun and M.J. Bevan. Defective CD8 T cell memory following acute infection without CD4 T cell help. *Science* **300**:339–342 (2003).

25. J.C. Sun, M.A. Williams and M.J. Bevan. CD4+ T cells are required for the maintenance, not programming, of memory CD8+ T cells after acute infection. *Nat Immunol* **5**:927–933 (2004).

26. K. Murali-Krishna, J.D. Altman, M. Suresh, D.J. Sourdive, A.J. Zajac, J.D. Miller, J. Slansky and R. Ahmed. Counting antigen-specific CD8 T cells: a reevaluation of bystander activation during viral infection. *Immunity* **8**:177–187 (1998).

27. R.G. van der Most, K. Murali-Krishna, J.G. Lanier, E.J. Wherry, M.T. Puglielli, J.N. Blattman, A. Sette and R. Ahmed. Changing immunodominance patterns in antiviral CD8 T-cell responses after loss of epitope presentation or chronic antigenic stimulation. *Virology* **315**:93–102 (2003).

28. E.J. Wherry, J.N. Blattman, K. Murali-Krishna, R. van der Most and R. Ahmed. Viral persistence alters CD8 T-cell immunodominance and tissue distribution and results in distinct stages of functional impairment. *J Virol* **77**:4911–4927 (2003).

29. K. Tewari, J. Sacha, X. Gao and M. Suresh. Effect of chronic viral infection on epitope selection, cytokine production, and surface phenotype of CD8 T cells and the role of IFN-gamma receptor in immune regulation. *J Immunol* **172**:1491–1500 (2004).

30. W. Chen, L.C. Anton, J.R. Bennink and J.W. Yewdell. Dissecting the multifactorial causes of immunodominance in class I-restricted T cell responses to viruses. *Immunity* **12**:83–93 (2000).

31. P.G. Stevenson, G.T. Belz, J.D. Altman and P.C. Doherty. Changing patterns of dominance in the CD8+ T cell response during acute and persistent murine gammaherpesvirus infection. *Eur J Immunol* **29**:1059–1067 (1999).

32. C.C. Bergmann, J.D. Altman, D. Hinton and S.A. Stohlman. Inverted immunodominance and impaired cytolytic function of CD8+ T cells during viral persistence in the central nervous system. *J Immunol* **163**:3379–3387 (1999).

33. C.C. Kemball, E.D. Lee, V. Vezys, T.C. Pearson, C.P. Larsen and A.E. Lukacher. Late priming and variability of epitope-specific CD8+ T cell responses during a persistent virus infection. *J Immunol* **174**:7950–7960 (2005).

34. N.M. Steven, A.M. Leese, N.E. Annels, S.P. Lee and A.B. Rickinson. Epitope focusing in the primary cytotoxic T cell response to Epstein-Barr virus and its relationship to T cell memory. *J Exp Med* **184**:1801–1813 (1996).

35. P.J. Goulder, A.K. Sewell, D.G. Lalloo, D.A. Price, J.A. Whelan, J. Evans, G.P. Taylor, G. Luzzi, P. Giangrande, R.E. Phillips and A.J. McMichael. Patterns of immunodominance in HIV-1-specific cytotoxic T lymphocyte responses in two human histocompatibility leukocyte antigens (HLA)-identical siblings with HLA-A*0201 are influenced by epitope mutation. *J Exp Med* **185**:1423–1433 (1997).

36. M.F. Bachmann, M. Barner, A. Viola and M. Kopf. Distinct kinetics of cytokine production and cytolysis in effector and memory T cells after viral infection. *Eur J Immunol* **29**:291–299 (1999).

37. B.J. Swanson, M. Murakami, T.C. Mitchell, J. Kappler and P. Marrack. RANTES production by memory phenotype T cells is controlled by a posttranscriptional, TCR-dependent process. *Immunity* **17**:605–615 (2002).

38. H. Veiga-Fernandes, U. Walter, C. Bourgeois, A. McLean and B. Rocha. Response of naive and memory CD8+ T cells to antigen stimulation in vivo. *Nat Immunol* **1**:47–53 (2000).

39. D.L. Barber, E.J. Wherry and R. Ahmed. Cutting edge: rapid in vivo killing by memory CD8 T cells. *J Immunol* **171**:27–31 (2003).

40. R.M. Welsh. Assessing CD8 T cell number and dysfunction in the presence of antigen. *J Exp Med* **193**:F19–F22 (2001).

41. E.J. Wherry, D.L. Barber, S.M. Kaech, J.N. Blattman and R. Ahmed. Antigen-independent memory CD8 T cells do not develop during chronic viral infection. *Proc Natl Acad Sci USA* **101**:16004–16009 (2004).

42. M.J. Fuller, A. Khanolkar, A.E. Tebo and A.J. Zajac. Maintenance, loss, and resurgence of T cell responses during acute, protracted, and chronic viral infections. *J Immunol* **172**:4204–4214 (2004).

43. A.M. Byers, C.C. Kemball, J.M. Moser and A.E. Lukacher. Cutting edge: rapid in vivo CTL activity by polyoma virus-specific effector and memory CD8+ T cells. *J Immunol* **171**:17–21 (2003).

44. E.J. Wherry, M.J. McElhaugh and L.C. Eisenlohr. Generation of CD8(+) T cell memory in response to low, high, and excessive levels of epitope. *J Immunol* **168**:4455–4461 (2002).

45. P.A. Goepfert, A. Bansal, B.H. Edwards, G.D. Ritter Jr., I. Tellez, S.A. McPherson, S. Sabbaj and M.J. Mulligan. A significant number of human immunodeficiency virus epitope-specific cytotoxic T lymphocytes detected by tetramer binding do not produce gamma interferon. *J Virol* **74**:10249–10255 (2000).

46. P. Shankar, M. Russo, B. Harnisch, M. Patterson, P. Skolnik and J. Lieberman. Impaired function of circulating HIV-specific CD8(+) T cells in chronic human immunodeficiency virus infection. *Blood* **96**:3094–3101 (2000).

47. S. Kostense, G.S. Ogg, E.H. Manting, G. Gillespie, J. Joling, K. Vandenberghe, E.Z. Veenhof, D. van Baarle, S. Jurriaans, M.R. Klein and F. Miedema. High viral burden in the presence of major HIV-specific CD8(+) T cell expansions: evidence for impaired CTL effector function. *Eur J Immunol* **31**:677–686 (2001).

48. N.H. Gruener, F. Lechner, M.C. Jung, H. Diepolder, T. Gerlach, G. Lauer, B. Walker, J. Sullivan, R. Phillips, G.R. Pape and P. Klenerman. Sustained dysfunction of antiviral CD8+ T lymphocytes after infection with hepatitis C virus. *J Virol* **75**:5550–5558 (2001).

49. H. Wedemeyer, X.S. He, M. Nascimbeni, A.R. Davis, H.B. Greenberg, J.H. Hoofnagle, T.J. Liang, H. Alter and B. Rehermann. Impaired effector function of hepatitis C virus-specific CD8+ T cells in chronic hepatitis C virus infection. *J Immunol* **169**:3447–3458 (2002).

50. V. Appay, D.F. Nixon, S.M. Donahoe, G.M. Gillespie, T. Dong, A. King, G.S. Ogg, H.M. Spiegel, C. Conlon, C.A. Spina, D.V. Havlir, D.D. Richman, A. Waters, P. Easterbrook, A.J. McMichael and S.L. Rowland-Jones. HIV-specific CD8(+) T cells

produce antiviral cytokines but are impaired in cytolytic function. *J Exp Med* **192**:63–75 (2000).

51. L.E. Gamadia, I.J. ten Berge, L.J. Picker and R.A. van Lier. Skewed maturation of virus-specific CTLs? *Nat Immunol* **3**:203 (2002).
52. L.E. Gamadia, E.B. Remmerswaal, J.F. Weel, F. Bemelman, R.A. van Lier and I.J. ten Berge. Primary immune responses to human CMV: a critical role for IFN-gamma-producing CD4+ T cells in protection against CMV disease. *Blood* **101**:2686–2692 (2003).
53. E.M. van Leeuwen, L.E. Gamadia, P.A. Baars, E.B. Remmerswaal, I.J. ten Berge and R.A. van Lier. Proliferation requirements of cytomegalovirus-specific, effector-type human CD8+ T cells. *J Immunol* **169**:5838–5843 (2002).
54. E.M. van Leeuwen, G.J. de Bree, E.B. Remmerswaal, S.L. Yong, K. Tesselaar, I.J. ten Berge and R.A. van Lier. IL-7 receptor alpha chain expression distinguishes functional subsets of virus-specific human CD8+ T cells. *Blood* **106**:2091–2098 (2005).
55. R.M. Welsh, L.K. Selin and E. Szomolanyi-Tsuda. Immunological memory to viral infections. *Annu Rev Immunol* **22**:711–743 (2004).
56. E.J. Wherry, V. Teichgraber, T.C. Becker, D. Masopust, S.M. Kaech, R. Antia, U.H. von Andrian and R. Ahmed. Lineage relationship and protective immunity of memory CD8 T cell subsets. *Nat Immunol* **4**:225–234 (2003).
57. A.L. Marzo, K.D. Klonowski, B.A. Le, P, Borrow, D.F. Tough and L. Lefrancois. Initial T cell frequency dictates memory CD8+ T cell lineage commitment. *Nat Immunol* **6**:793–799 (2005).
58. M.J. Fuller, D.A. Hildeman, S. Sabbaj, D.E. Gaddis, A.E. Tebo, L. Shang, P.A. Goepfert and A.J. Zajac. Cutting edge: emergence of CD127high functionally competent memory T cells is compromised by high viral loads and inadequate T cell help. *J Immunol* **174**:5926–5930 (2005).
59. K.S. Lang, M. Recher, A.A. Navarini, N.L. Harris, M. Lohning, T. Junt, H.C. Probst, H. Hengartner and R.M. Zinkernagel. Inverse correlation between IL-7 receptor expression and CD8 T cell exhaustion during persistent antigen stimulation. *Eur J Immunol* **35**:738–745 (2005).
60. R.A. van Lier, I.J. ten Berge and L.E. Gamadia. Human CD8(+) T-cell differentiation in response to viruses. *Nat Rev Immunol* **3**:931–939 (2003).
61. X.S. He, K. Mahmood, H.T. Maecker, T.H. Holmes, G.W. Kemble, A.M. Arvin and H.B. Greenberg. Analysis of the frequencies and of the memory T cell phenotypes of human CD8+ T cells specific for influenza A viruses. *J Infect Dis* **187**:1075–1084 (2003).
62. G.J. de Bree, J. Heidema, E.M. van Leeuwen, G.M. van Bleek, R.E. Jonkers, H.M. Jansen, R.A. van Lier and T.A. Out. Respiratory syncytial virus-specific CD8+ memory T cell responses in elderly persons. *J Infect Dis* **191**:1710–1718 (2005).
63. V. Appay and S.L. Rowland-Jones. Lessons from the study of T-cell differentiation in persistent human virus infection. *Semin Immunol* **16**:205–212 (2004).
64. X.S. He, B. Rehermann, F.X. Lopez-Labrador, J. Boisvert, R. Cheung, J. Mumm, H. Wedemeyer, M. Berenguer, T.L. Wright, M.M. Davis and H.B. Greenberg. Quantitative analysis of hepatitis C virus-specific CD8(+) T cells in peripheral blood and liver using peptide-MHC tetramers. *Proc Natl Acad Sci USA* **96**:5692–5697 (1999).
65. G.S. Ogg, S. Kostense, M.R. Klein, S. Jurriaans, D. Hamann, A.J. McMichael and F. Miedema. Longitudinal phenotypic analysis of human immunodeficiency virus type

1-specific cytotoxic T lymphocytes: correlation with disease progression. *J Virol* **73**:9153–9160 (1999).

66. M.D. Catalina, J.L. Sullivan, R.M. Brody and K. Luzuriaga. Phenotypic and functional heterogeneity of EBV epitope-specific CD8+ T cells. *J Immunol* **168**:4184–4191 (2002).
67. V. Appay, P.R. Dunbar, M. Callan, P. Klenerman, G.M. Gillespie, L. Papagno, G.S. Ogg, A. King, F. Lechner, C.A. Spina, S. Little, D.V. Havlir, D.D. Richman, N. Gruener, G. Pape, A. Waters, P. Easterbrook, M. Salio, V. Cerundolo, A.J. McMichael and S.L. Rowland-Jones. Memory CD8+ T cells vary in differentiation phenotype in different persistent virus infections. *Nat Med* **8**:379–385 (2002).
68. G. Chen, P. Shankar, C. Lange, H. Valdez, P.R. Skolnik, L. Wu, N. Manjunath and J. Lieberman. CD8 T cells specific for human immunodeficiency virus, Epstein-Barr virus, and cytomegalovirus lack molecules for homing to lymphoid sites of infection. *Blood* **98**:156–164 (2001).
69. F. Kern, E. Khatamzas, I. Surel, C. Frommel, P. Reinke, S.L. Waldrop, L.J. Picker and H.D. Volk. Distribution of human CMV-specific memory T cells among the CD8pos subsets defined by CD57, CD27, and CD45 isoforms. *Eur J Immunol* **29**:2908–2915 (1999).
70. L.E. Gamadia, R.J. Rentenaar, P.A. Baars, E.B. Remmerswaal, S. Surachno, J.F. Weel, M. Toebes, T.N. Schumacher, I.J. ten Berge and R.A. van Lier. Differentiation of cytomegalovirus-specific CD8(+) T cells in healthy and immunosuppressed virus carriers. *Blood* **98**:754–761 (2001).
71. M.R. Wills, G. Okecha, M.P. Weekes, M.K. Gandhi, P.J. Sissons and A.J. Carmichael. Identification of naive or antigen-experienced human CD8(+) T cells by expression of costimulation and chemokine receptors: analysis of the human cytomegalovirus-specific CD8(+) T cell response. *J Immunol* **168**:5455–5464 (2002).
72. J.K. Sandberg, N.M. Fast and D.F. Nixon. Functional heterogeneity of cytokines and cytolytic effector molecules in human CD8+ T lymphocytes. *J Immunol* **167**:181–187 (2001).
73. G.J. de Bree, E.M. van Leeuwen, T.A. Out, H.M. Jansen, R.E. Jonkers and R.A. van Lier. Selective accumulation of differentiated CD8+ T cells specific for respiratory viruses in the human lung. *J Exp Med* **202**:1433–1442 (2005).
74. L.E. Gamadia, E.M. van Leeuwen, E.B. Remmerswaal, S.L. Yong, S. Surachno, P.M. Wertheim-van Dillen, I.J. ten Berge and R.A. van Lier. The size and phenotype of virus-specific T cell populations is determined by repetitive antigenic stimulation and environmental cytokines. *J Immunol* **172**:6107–6114 (2004).
75. M. Paiardini, B. Cervasi, H. Albrecht, A. Muthukumar, R. Dunham, S. Gordon, H. Radziewicz, G. Piedimonte, M. Magnani, M. Montroni, S.M. Kaech, A. Weintrob, J.D. Altman, D.L. Sodora, M.B. Feinberg and G. Silvestri. Loss of CD127 expression defines an expansion of effector CD8+ T cells in HIV-infected individuals. *J Immunol* **174**:2900–2909 (2005).
76. L.L. Lau, B.D. Jamieson, T. Somasundaram and R. Ahmed. Cytotoxic T-cell memory without antigen. *Nature* **369**:648–652 (1994).
77. S. Hou, L. Hyland, K.W. Ryan, A. Portner and P.C. Doherty. Virus-specific CD8+ T-cell memory determined by clonal burst size. *Nature* **369**:652–654 (1994).
78. K.S. Schluns, W.C. Kieper, S.C. Jameson and L. Lefrancois. Interleukin-7 mediates the homeostasis of naive and memory CD8 T cells in vivo. *Nat Immunol* **1**:426–432 (2000).

79. J.T. Tan, B. Ernst, W.C. Kieper, E. LeRoy, J. Sprent and C.D. Surh. Interleukin (IL)-15 and IL-7 jointly regulate homeostatic proliferation of memory phenotype CD8+ cells but are not required for memory phenotype CD4+ cells. *J Exp Med* **195**:1523–1532 (2002).

80. M. Prlic, L. Lefrancois and S.C. Jameson. Multiple choices: regulation of memory CD8 T cell generation and homeostasis by interleukin (IL)-7 and IL-15. *J Exp Med* **195**:F49–F52 (2002).

81. T.C. Becker, E.J. Wherry, D. Boone, K. Murali-Krishna, R. Antia, A. Ma and R. Ahmed. Interleukin 15 is required for proliferative renewal of virus-specific memory CD8 T cells. *J Exp Med* **195**:1541–1548 (2002).

82. J.J. Obar, S.G. Crist, E.K. Leung and E.J. Usherwood. IL-15-independent proliferative renewal of memory CD8+ T cells in latent gammaherpesvirus infection. *J Immunol* **173**:2705–2714 (2004).

83. R. Dudani, Y. Chapdelaine, H.H. Faassen, D.K. Smith, H. Shen, L. Krishnan and S. Sad. Multiple mechanisms compensate to enhance tumor-protective CD8(+) T cell response in the long-term despite poor CD8(+) T cell priming initially: comparison between an acute versus a chronic intracellular bacterium expressing a model antigen. *J Immunol* **168**:5737–5745 (2002).

84. Y. Belkaid, C.A. Piccirillo, S. Mendez, E.M. Shevach and D.L. Sacks. CD4+CD25+ regulatory T cells control Leishmania major persistence and immunity. *Nature* **420**:502–507 (2002).

85. J.H. Park, Q. Yu, B. Erman, J.S. Appelbaum, D. Montoya-Durango, H.L. Grimes and A. Singer. Suppression of IL7Ralpha transcription by IL-7 and other prosurvival cytokines: a novel mechanism for maximizing IL-7-dependent T cell survival. *Immunity* **21**:289–302 (2004).

86. B.M. Foxwell, D.A. Taylor-Fishwick, J.L. Simon, T.H. Page and M. Londei. Activation induced changes in expression and structure of the IL-7 receptor on human T cells. *Int Immunol* **4**:277–282 (1992).

87. C. Carini, M.F. McLane, K.H. Mayer and M. Essex. Dysregulation of interleukin-7 receptor may generate loss of cytotoxic T cell response in human immunodeficiency virus type 1 infection. *Eur J Immunol* **24**:2927–2934 (1994).

88. P.A. MacPherson, C. Fex, J. Sanchez-Dardon, N. Hawley-Foss and J.B. Angel. Interleukin-7 receptor expression on CD8(+) T cells is reduced in HIV infection and partially restored with effective antiretroviral therapy. *J Acquir Immune Defic Syndr* **28**:454–457 (2001).

89. C. Mussini, M. Pinti, V. Borghi, M. Nasi, G. Amorico, E. Monterastelli, L. Moretti, L. Troiano, R. Esposito and A. Cossarizza. Features of 'CD4-exploders', HIV-positive patients with an optimal immune reconstitution after potent antiretroviral therapy. *AIDS* **16**:1609–1616 (2002).

10

GERMINAL CENTER-DERIVED B CELL MEMORY

Craig P. Chappell and Joshy Jacob

1. INTRODUCTION

B cell memory is characterized by persistent levels of Ag-specific serum antibody (Ab) following immunization and the ability to rapidly produce Ab upon secondary Ag exposure[1]. During primary immune responses in mammals, B cell activation occurs within or at the border of T cell-rich periarteriolar lymphoid sheaths following cognate interaction with CD4 T_H cells[2,3]. This interaction induces B lymphocytes to differentiate into either foci of antibody-forming cells (AFC) found in red pulp and follicular borders or germinal center (GC) B cells located within secondary follicles[2-4]. The primary foci response in mice is relatively short lived (5–10 days); in contrast, the GC response is a more sustained program of cellular differentiation in which extensive B cell proliferation, somatic hypermutation of IgV gene segments, and memory cell selection occurs[4-8]. While GC B cells are readily identified using cell surface attributes, no such markers have been identified on murine memory B cells. The inability to reliably identify memory B cells has hampered the study of memory B cell development and differentiation.

Therefore, we sought to develop a mouse model that would identify Ag-specific memory B cells in a manner that did not rely on Ag-labeling techniques and would simultaneously allow for recovery of the cells following their identification. We generated double transgenic mice in which a tractable marker, β-galactosidase (β-gal), was indelibly expressed in germinal center B cells using a cre-lox recombination strategy. The permanent nature of β-gal expression allowed us to identify and isolate Ag-specific B cells long after immunization with the model antigen, (4-hydroxy-3-nitrophenyl)acetyl chicken gamma globulin (NPCG). Our results show that β-gal+ B cells persisted into the immune phase

Emory University, Department of Microbiology and Immunology, Atlanta, GA, 30329. (tel) 404-727-7919 (fax) 404-727-8199. jjacob3@emory.edu

of the response and exclusively contained hypermutated λ_1 V regions. Importantly, adoptive transfer experiments demonstrated that β-gal$^+$ B cells from immune mice produced >100-fold more AFC than their β-gal$^-$ counterparts following Ag challenge, demonstrating both a hallmark of B cell memory and Ag specificity.

2. MATERIALS AND METHODS

2.1. Mice and Immunizations

ROSA26R and congenic Ly5.1 mice were purchased from Jackson Laboratories, (Maine). Germinal center-cre (GCC) transgenic founders (lines GCC1.6, GCC816, and GCC158) were derived in the C57BL/6 background at Rockefeller University (New York) (GCC816 and GCC1.6) or Emory University Transgenic Core Facility (Atlanta, GA) (GCC158). Primers used for genotyping ROSA26R mice via PCR have been described[9]. Primers for cre-specific PCR are as follows: cre-forward 5'-ACACCCTGTTACGTAT AGCCGAAA; cre-reverse 5'-TATTCGGATCATCAGCT ACACCAG. All mice were housed under specific pathogen-free conditions at the Emory Vaccine Center. The succinimide ester of the hapten NP (Cambridge Research Biochemicals Ltd., Gadbrook Park, Northwich, Chesire, UK) was coupled to chicken gamma globulin (Sigma, St. Louis, MO), and the NP:CG molar ratios were determined via spectrophotometry. We used $NP_{14}CG$ and $NP_{22}CG$ for these studies. Primary immune responses were induced by a single intraperitoneal (i.p.) injection of 50-μg alum-precipitated NPCG. Secondary responses were induced via tail vein injection of 20-μg NPCG in PBS. All studies were approved by Emory University's Institutional Animal Care and Use Committee (IACUC).

2.2. β-Galactosidase Detection, Antibodies, and Flow Cytometry

For detection of β-galactosidase activity, 1-10 x 10^7 RBC-cleared splenocytes were washed once with 1X PBS, re-suspended in 0.2–0.5ml PBS, and warmed to 37°C. Next, an equal volume of 250-μM fluorescein-di-β-galactopyranoside (FDG) (Molecular Probes, Eugene, OR) in dH_2O (pre-warmed to 37°C) was added to the cells, mixed, and incubated at 37°C for 2 minutes in an air incubator. The loading was quenched by addition of 10 volumes cold PBS. For antibody staining, $1–2 \times 10^6$ splenocytes were next incubated on ice for 20 minutes with optimal concentrations of anti-B220-allophycocyanin (Caltag), CD38-PE (JC11), CD95-PE, GL-7-biotin, CD138-PE (clone 281.2), CD4-PerCP, CD8-PerCP, CD8-PE, CXCR5-PE, CD19-PE, CD11c-PE, F4/80-PE, GR1-PE (all from BD Biosciences), or PNA-biotin (Sigma). Biotinylated reagents were detected with 5-μg/ml streptavidin-PerCP (BD Biosciences). Following washes, cells were kept cold and acquired on a FACSCalibur (BD Biosciences) running CellQuest software. Data were analyzed using FlowJo software (TreeStar Inc.).

2.3. ELISPOT Assay

ELISPOT assays were carried out as described[10], with the exception that 96-well nitrocellulose plates (Millipore) were coated overnight with 20-μg/ml NPCG in 50-μl PBS. Spots were visualized using an ELISPOT reader (Cellular Technologies Ltd., Cleveland, OH) and counted manually.

2.4. Cell Sorting and Adoptive Transfers

For cell-sorting experiments, RBC-cleared splenocytes were loaded with FDG and stained with antibodies as described above. For isolation of memory B cells, B220$^+$Synd-1$^-$ cells from immune (day 60–120 p.i.) GCCxR26R mice were sorted into β-gal$^+$ and β-gal$^-$ fractions. CD4$^+$ T cells from immune (>35 days p.i.) cre-negative GCCxR26R mice were negatively sorted by excluding all cells that expressed CD19, CD8, Synd-1, CD11c, F4/80, and GR1. Sorted cell purities ranged from 85 to 98%. For adoptive transfers, sorted splenocytes were combined and administered to the tail vein of RAG2$^{-/-}$ recipients. 48 hr following transfer, recipients were challenged i.p. with 20-μg NPCG in PBS.

2.5. PCR and DNA Sequencing

Semi-nested PCR (Roche High-Fidelity FastStart PCR Kit) was used to amplify the λ_1 light chain V–J region using primer sequences previously reported[11]. PCR products were purified (Qiagen PCR Clean Kit), digested overnight with *Bam*HI and *Hin*dIII (NEB), agarose purified (Qiagen), subcloned into pBluescript$^{+/-}$ (Invitrogen), and used for transforming TOP10F' bacteria (Invitrogen). Templates for sequencing were generated by colony PCR using M13-F and M13-R primers. Sequencing reactions (DYEnamic ET Dye Terminator, Amersham Biosciences) were acquired on a MegaBACE 1000 (Amersham Biosciences) at the Emory Vaccine Center or, alternatively, by Macrogen Inc. (Seoul, South Korea). Sequencing results were analyzed with EditView (ABI Prism) and MacClade software (Sinauer Associates Inc, Sunderland, MD).

2.6. Statistics

Student's *t* test was used to generate all statistical values stated. For statistical designations * denotes $p < 0.05$, ** denotes $p < 0.01$, and *** denotes $p < 0.001$.

3. RESULTS

3.1. Generation of Germinal Center-Cre Transgenic Mice

To develop a mouse model to study B cell memory, we generated three individual lines of transgenic mice containing the −1.23-kb *Sma*I-truncated MHC class II I-E$_\alpha^d$ promotor driving the expression of *cre recombinase* (termed germinal center-cre, GCC). Previous transgenic studies using the –1.23-kb I-E$_\alpha^d$ promotor demonstrated that gene expression from this promotor was restricted in B lineage cells to the germinal center stage of B cell development[12,13]. GCC mice were subsequently bred to the ROSA26R (R26R) *cre*-reporter strain, in which constitutive expression of β-galactosidase (β-gal) occurs upon *cre*-mediated recombination of the ROSA locus[9] (Figure 1). GCCxR26R offspring (termed GCCxR26R) developed normally and contained average numbers of lymphocytes in both spleen and lymph nodes (data not shown) compared to single transgenic littermate controls.

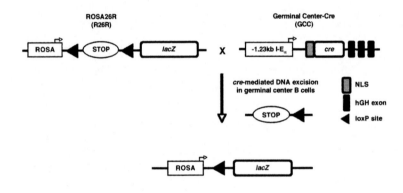

Figure 1. Cre recombinase expression driven by the truncated –1.23 kb I-E$_\alpha^d$ promotor in germinal center B cells leads to excision of intervening transcriptional and translational stop signals via recognition of flanking loxP sites. Removal of the block in transcription allows for β-gal expression from the constitutive ROSA promotor. NLS, nuclear localization signal; hGH, human growth hormone.

3.2. Splenic β-gal Expression Is Induced upon Immunization

As shown in Figure 2a, naive GCCxR26R mice displayed little β-gal expression or PNA binding among splenic B cells. However, 16 days following immunization with NPCG, both PNA$^+$ and β-gal$^+$ B cells were detected in the spleens of GCCxR26R mice while single transgenic littermate controls displayed PNA$^+$ B cells only. These results demonstrate that β-gal expression is induced following immunization and is associated with the development of GC. To investigate the kinetics of β-gal expression following immunization, we immunized cohorts of GCCxR26R mice and single transgenic littermate controls with NPCG and analyzed splenic B cells for β-gal expression and PNA binding by flow cytometry at various time points following immunization. We found that β-gal$^+$B220$^+$ B

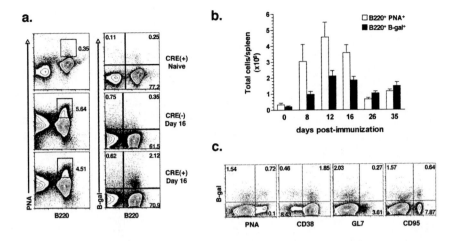

Figure 2. (a) Splenocytes from naive or immunized (day 16 NPCG) GCCxR26R and single transgenic control mice were analyzed for β-gal and PNA expression. Flow cytometry plots show that significant β-gal expression is found only after immunization in double transgenic GCCxR26R mice. **(b)** Mean (±SEM) total numbers of PNA⁺ and β-gal⁺ B cells per spleen following NPCG immunization are graphed. $n = 3$–7 per time point. **(c)** Flow cytometry plots show GC activation markers among gated B220⁺ B cells taken from the spleens of GCCxR26R mice 10 days after NPCG immunization. Shown are representative plots from 5 individual mice.

cells were evident in the spleen 8 days following immunization of GCCxR26R mice and increased in number until day 12, accounting for $2.1 \pm 0.38\%$ of total splenocytes ($3.5 \pm 0.54\%$ of B220⁺ cells) at their peak (Figure 2b). This was followed by a modest contraction phase that reached steady-state levels by day 26 post-immunization. The kinetics of β-gal expression was similar to PNA⁺ GC B cell development; however, the frequency and total number of β-gal⁺ B cells per spleen was consistently less than PNA⁺ B cells during the early to intermediate stages of the primary response. During the immune phase of the response (>5 weeks p.i.), the number of β-gal⁺ B cells was higher than PNA⁺ B cells, suggesting the development of β-gal⁺ memory B cells over time (data not shown). Immunized single transgenic littermate controls failed to express β-gal at all time points examined (data not shown).

3.3. β-Galactosidase Expression Does not Mark All GC B Cells

Next, we determined the extent of correlation between β-gal expression and markers associated with GC B cells following immunization with NPCG[14]. Flow cytometry analysis of β-gal⁺ B cells from the spleens of GCCxR26R mice immunized 10 days previously showed that of $22.7 \pm 1.4\%$ bound PNA, $13.1 \pm 5.3\%$ were CD38ᵗᵒ, and expression of GL7 and CD95 was detected in $16.0 \pm$

3.8% and 20.3 ± 4.7% of β-gal⁺ B cells, respectively (Figure 2c). This finding was not unique to the time point analyzed. Analysis on days 8, 16, 26, and 35 post-immunization showed that, while the frequency of both PNA⁺ and β-gal⁺ B cells rose and fell over time, the frequency of PNA⁺ B cells within the β-gal⁺ B cell fraction did not significantly deviate during primary response (data not shown). In addition, β-gal⁺B220⁺ cells were uniformly positive for CD19 and CXCR5 expression, and did not express CD4 or CD8 (data not shown). Finally, ≥90% of β-gal⁺ cells were negative for syndecan-1 (Synd-1) expression 10 days p.i., indicating that primary AFC were not marked by β-gal expression (data not shown). Together, these results demonstrate that β-gal does not mark all PNA⁺ GC B cells.

3.4. β-gal⁺ GC B Cells Contain Mutated λ_1 V Regions

The phenotype of β-gal⁺ B cells was unexpected and suggested that many did not participate in GC reactions. To better understand the phenotype of the β-gal⁺ B cell populations, we determined which populations of β-gal⁺ B cells underwent hypermutation of their rearranged λ_1 V region loci following immunization. We immunized cohorts of GCCxR26R mice with NPCG, harvested and pooled their spleens, and sorted B220⁺ B cells with β-gal⁻PNA⁺, β-gal⁺PNA⁺, and β-gal⁺PNA⁻ phenotypes 8 or 15 days following immunization ($n = 3$ per time point). Due to the predominant use of λ_1 L chains in mice of the Igh^b allotype during the primary response to NP, this strategy allowed us to selectively amplify V regions of NP-specific B cells[15,16].

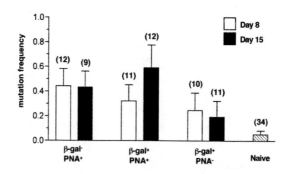

Figure 3. λ_1 V region mutation frequencies of the indicated GC B cell populations are graphed for day 8 and day 15 post-primary immunization with NPCG. The numbers in parentheses indicate the number of clones analyzed for each population. Mutation frequencies were calculated by dividing the total number of mutations in each population by the total number of base pairs sequenced and the result multiplied by 100.

The mutation frequency of each population is shown in Figure 3. 60 and 82% of β-gal$^+$PNA$^-$ B cells at days 8 and 15 p.i., respectively, did not contain mutations. This yielded a low frequency of mutation at both time points (0.248–0.193), which suggested either these cells did not participate in GC or did not survive long enough within GC to accumulate significant mutations. In contrast, 58% of sequences recovered from β-gal$^-$PNA$^+$ B cells at day 8 p.i. were mutated, and this frequency rose to 67% by day 15 p.i. A high frequency (73–82%) of mutations in β-gal$^-$PNA$^+$ B cells gave rise to amino acid replacements, and the percentage of mutations that fell within CDR rose from 47% on day 8 to 73% by day 15. The mutation frequency of β-gal$^-$PNA$^+$ B cells was ~2-fold higher than β-gal$^+$PNA$^-$ B cells and ~10-fold higher than that obtained from naive B cells. Similar to β-gal$^-$PNA$^+$ B cells, the frequency of mutated β-gal$^+$PNA$^+$ B cells increased from day 8 to day 15 (45–75%). Likewise, the mutation frequency of these cells also increased from day 8 to day 15 p.i. The pattern of mutations within β-gal$^+$PNA$^+$ B cells was highly indicative of Ag-driven selection; 80–100% of mutations fell within CDR and gave rise to amino acid replacements 80–90% of the time. The presence and pattern of somatic hypermutation within rearranged λ_1 V regions demonstrates that β-gal$^+$ B cells with a GC phenotype (PNA$^+$) participated in GC reactions and were under Ag selection pressures.

3.5. Hypermutated β-gal$^+$ Memory B Cells Transfer Ag Recall Responses

To determine whether the β-gal$^+$ B cell population in immune GCCxR26R mice represented authentic memory B cells, we analyzed rearranged λ_1 V gene segments from β-gal$^+$ and β-gal$^-$ B cells taken from immune GCCxR26R mice for evidence of hypermutation and Ag-driven selection. GCCxR26R mice were immunized with NPCG and sacrificed 60 days later, or, alternatively, boosted with soluble Ag and rested for 50 days before sacrifice. B220$^+$ cells were then sorted according to β-gal expression and sequenced as above. 79% (38/48) of clones from B220$^+$β-gal$^+$ B cells displayed up to 5 mutations per sequence. On the other hand, B220$^+$β-gal$^-$ B cells displayed very few mutations; 80% (33/41) were unmutated, while the remaining 20% contained a single mutation each. The mutation frequency of rearranged λ_1 loci in β-gal$^-$ B cells did not significantly differ from that observed in naive splenocytes (0.069 vs. 0.052, $p = 0.65$) (Figure 4a). In contrast, the mutation frequency in β-gal$^+$ memory B cells was 8.1-fold higher than β-gal$^-$ B cells (0.560 vs. 0.069, $p < 0.001$), and 10.7-fold higher than splenocytes taken from a naive GCCxR26R control (0.560 vs. 0.052, $p < 0.001$). 72% of unique mutations from β-gal$^+$ B cells yielded amino acid changes, and 58% of these fell within CDR, a pattern indicative of Ag-driven selection. These results show that λ_1 V region mutations are found almost exclusively within the β-gal$^+$ B cell pool in immune GCCxR26R mice.

Next, we determined whether β-gal$^+$ memory B cells were capable of transferring memory responses to naive mice. We performed adoptive transfers in which 5×10^4 purified β-gal$^+$ or β-gal$^-$ B cells from immune GCCxR26R mice (day 60–120 p.i.) were transferred to immunodeficient RAG2$^{-/-}$ hosts (schematic in Figure 4a). To provide a source of T cell help, each mouse received ~1 × 10^6 CD4$^+$ T cells from NPCG-immune (>day 60 p.i.) R26R mice. In addition, each recipient also received 3 × 10^6 splenocytes from naive Ly5.1 mice to aid in lymphoid follicle formation in the RAG2$^{-/-}$ hosts. To control for possible memory B cell contamination in the CD4$^+$ T cell fraction, one cohort received only CD4$^+$ T cells from immune mice plus the naive splenocyte fraction. Additionally, 2 groups of mice received unfractionated splenocytes (5 × 10^6) from immune GCCxR26R mice in which one group was given Ag and the other was left unchallenged. The recipients were challenged i.p. with 20-μg soluble NPCG 48 hr following cell transfer. We analyzed recipient spleens 5 days following Ag challenge for the presence of IgG-secreting NPCG-specific AFC by ELISPOT (Figure 4b).

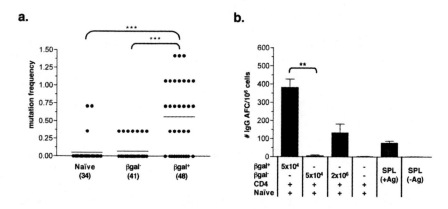

Figure 4. (a) The frequency of mutations within rearranged λ, V regions is plotted for individual clones derived from the indicated population of B cells taken from naive or immune (>day 60 p.i.) GCCxR26R mice. Mutation frequencies were calculated as in Figure 2. The number of clones analyzed from each population are indicated in parentheses. (b) ELISPOT assay 5 days following Ag challenge of RAG2$^{-/-}$ recipients showing the number of splenic AFC obtained from the transfer of the indicated number of the cell type listed. Plotted are the combined results from 2 of 3 representative experiments. SPL = non-fractionated splenocyte transfers.

As expected, Ag-challenged recipients that received non-fractionated splenocytes from immune GCCxR26R mice contained elevated numbers of splenic AFC capable of secreting NPCG-specific IgG (76 ± 10 spots/10^6 cells). We were unable to detect Ag-specific AFC from the spleens of either unimmunized recipients of non-fractionated splenocytes, nor recipients of CD4$^+$ T cells alone (<4 spots per 10^6 cells). In contrast, animals that received β-gal$^+$ memory B cells contained high numbers of AFC that secreted Ag-specific IgG (382 ± 45 per 10^6). AFC detection in β-gal$^-$ recipient spleens was virtually absent (5.8 ± 4.9 per 10^6). However, upon transfer of 40-fold higher numbers of β-gal$^-$ B cells from immune mice, a consistent AFC response was detected. This indicated that memory B cells were present in the β-gal⁻ fraction, albeit at a highly reduced (116-fold) frequency compared to β-gal⁺ B cells from immune mice.

4. CONCLUSION

The results presented in this report demonstrate the identification of germinal center-derived memory B cells in mice following immunization with a hapten–protein conjugate. In a novel mouse model system, we induced permanent β-galactosidase expression in GC B cells via *cre*-mediated recombination, which allowed us to detect both B cells participating in GC reactions and long-lived Ag-specific memory B cells that persisted long term following the initial antigenic stimulus. Our results showed that β-gal⁺ memory B cells contain three hallmark traits of B cell memory: (i) Ag specificity; (ii) hypermutated IgV gene segments; and (iii) the ability to transfer memory responses following adoptive transfer to naive hosts. This model is a unique demonstration of memory B cell identification using techniques that do not rely on Ag-labeling methods, and therefore should facilitate the study of B cell memory following multi-protein immunization as well as live pathogenic infection.

5. REFERENCES

1. D. Gray. Immunological memory. *Annu Rev Immunol* 11:49–77 (1993).
2. Y.J. Liu, J. Zhang, P.J. Lane, E.Y. Chan and I.C. MacLennan. Sites of specific B cell activation in primary and secondary responses to T cell-dependent and T cell-independent antigens. *Eur J Immunol* 21:2951–2962 (1991).
3. J. Jacob, R. Kassir and G. Kelsoe. In situ studies of the primary immune response to (4-hydroxy-3-nitrophenyl)acetyl, I: the architecture and dynamics of responding cell populations. *J Exp Med* 173:1165–1175 (1991).
4. J. Jacob and G. Kelsoe. In situ studies of the primary immune response to (4-hydroxy-3-nitrophenyl)acetyl, II: a common clonal origin for periarteriolar lymphoid sheath-associated foci and germinal centers. *J Exp Med* 176:679–687 (1992).
5. M.G. McHeyzer-Williams, M.J. McLean, P.A. Lalor and G.J. Nossal. Antigen-driven B cell differentiation in vivo. *J Exp Med* 178:295–307 (1993).
6. I.C. MacLennan and D. Gray. Antigen-driven selection of virgin and memory B cells. *Immunol Rev* 91:61–85 (1986).
7. J. Jacob, J. Przylepa, C. Miller and G. Kelsoe. In situ studies of the primary immune response to (4-hydroxy-3-nitrophenyl)acetyl, III: the kinetics of V region mutation and selection in germinal center B cells. *J Exp Med* 178:1293–1307 (1993).
8. S. Han, B. Zheng, J. Dal Porto and G. Kelsoe. In situ studies of the primary immune response to (4-hydroxy-3-nitrophenyl)acetyl, IV: affinity-dependent, antigen-driven B cell apoptosis in germinal centers as a mechanism for maintaining self-tolerance. *J Exp Med* 182:1635–1644 (1995).
9. P. Soriano. Generalized lacZ expression with the ROSA26 Cre reporter strain. *Nat Genet* 21:70–71 (1999).

10. M.K. Slifka and R. Ahmed. Limiting dilution analysis of virus-specific memory B cells by an ELISPOT assay. *J Immunol Methods* 199:37–46 (1996).

11. H. Jacobs, Y. Fukita, G.T. van der Horst, J. de Boer, G. Weeda, J. Essers, N. de Wind, B.P. Engelward, L. Samson, S. Verbeek, J.M. de Murcia, G. de Murcia, H. te Riele and K. Rajewsky. Hypermutation of immunoglobulin genes in memory B cells of DNA repair-deficient mice. *J Exp Med* 187:1735–1743 (1998).

12. W. van Ewijk, Y. Ron, J. Monaco, J. Kappler, P. Marrack, M. Le Meur, P. Gerlinger, B. Durand, C. Benoist and D. Mathis. Compartmentalization of MHC class II gene expression in transgenic mice. *Cell* 53:357–370 (1988).

13. L.C. Burkly, D. Lo, C. Cowing, R.D. Palmiter, R.L. Brinster, and R.A. Flavell. Selective expression of class II E alpha d gene in transgenic mice. *J Immunol* 142:2081–2088 (1989).

14. S.M. Shinall, M. Gonzalez-Fernandez, R.J. Noelle and T.J. Waldschmidt. Identification of murine germinal center B cell subsets defined by the expression of surface isotypes and differentiation antigens. *J Immunol* 164:5729–5738 (2000).

15. K. Karjalainen, B. Bang and O. Makela. Fine specificity and idiotypes of early antibodies against (4-hydroxy-3-nitrophenyl)acetyl (NP). *J Immunol* 125:313–317 (1980).

16. A. Cumano and K. Rajewsky. Structure of primary anti-(4-hydroxy-3-nitrophenyl)acetyl (NP) antibodies in normal and idiotypically suppressed C57BL/6 mice. *Eur J Immunol* 15:512–520 (1985).

11

CD28 AND CD27 COSTIMULATION OF CD8+ T CELLS: A STORY OF SURVIVAL

Douglas V. Dolfi and Peter D. Katsikis

1. INTRODUCTION

The role of costimulation in antiviral CD8+ T cells responses is becoming increasingly important as we try to develop adjuvant technologies for therapeutic or vaccine applications. Understanding how costimulation signals work and being able to harness their function to promote robust and protective immune responses is of particular interest. Much of current immunological research is addressing the many costimulatory molecules that are being discovered and characterized in order to elucidate the different mechanisms by which they work. It is becoming clear that multiple costimulation molecules are involved during the different phases of the CD8+ T cell response providing important proliferative and survival signals for these cells. The concept of T cell costimulation has evolved from the initial concept of the single CD28 second signal to an increasingly complex array of costimulation signals that involve multiple members of the B7:CD28 and TNFα/TNFR families. This review will focus on CD28 and CD27 costimulation and examine their involvement in the costimulation of CD8+ T cell responses and the role of such costimulation in the survival of activated, resting, and memory CD8+ T cells. We will also examine the importance of costimulatory-induced survival in antiviral CD8+ T cell responses.

2. CLASSICAL AND ALTERNATIVE COSTIMULATION

The idea that lymphocytes require additional signals to those provided by their antigen-specific receptors was first proposed in 1970 by Bretscher and Cohn in

Department of Microbiology and Immunology, Drexel University College of Medicine, Philadelphia PA 19129, USA. Address correspondence to: Peter D. Katsikis: peter.katsikis@drexelmed.edu. This work was supported by NIH grant R01 AI66215 awarded to PDK.

their two-signal theory of self/non-self discrimination[1]. The observation that
T cells become activated and proliferate in response to TCR stimulation only in
the presence of accessory cells, now known as APCs, was the first experimental
evidence that costimulation was critical for a T cell response to occur[2]. However, the nature of costimulation was not initially known and was thought to be
provided by soluble components expressed by the accessory cells, which is both
true and false, as discussed below. However, soluble factors are only part of the
story and are not sufficient in most cases to initiate proper effector T cell formation. When anti-CD28 antibody was discovered to replace the need for activated
splenocytes, the first molecule of what we now term costimulatory molecules
was identified. As opposed to the action of IL-2, which also induces strong proliferation of T cells and T cell clones, CD28 stimulation has the ability to prevent anergy of T cells stimulated through the TCR in the absence of accessory
cells[3]. Costimulation via CD28 also increases the ability of T cells to produce
IL-2 as well as other factors that increase proliferation. The interaction between
CD28 and its B7 family member ligands has become the hallmark of the two-
signal theory of T cell activation and the classical costimulatory pathway in immunology. Although CD28 is one of the most investigated molecules in immunological research, many facets of CD28 signaling are still unknown. The identification of additional novel members of the CD28 receptor family and their
respective B7 family ligands has revealed an ever-increasing complexity of
costimulation. To the known CD28 and CTLA-4 receptors, the new CD28 family members ICOS, PD-1, and BTLA have been added to the family[4,5]. On the
other hand, the B7 family of ligands is now comprised of B7-1 (CD80), B7-2
(CD86), ICOS ligand, PD-L1 (B7-H1), PD-L2 (B7-DC), B7-H3, and B7-H4
(B7x, B7-S1)[4]. The above ligands are expressed by antigen-presenting cells as
well as non-lymphoid tissues. Signaling through the CD28 family members provides both positive and negative signals that either augment, sustain, downregulate, or terminate T cell responses.

Similar roles have been identified for TNF family members and their respective receptors. The receptors CD40, Ox40, 4-1BB (CD137), TNFR I and II,
CD95/Fas, and CD27 and their respective ligands have all been shown to be
involved in the regulation of CD4+ and CD8+ T cell responses[6]. Thus the initial two-signal theory that involved a TCR signal and a single second signal that
is required for initiation of T cell proliferation has evolved to a much more complex concept of a TCR signal combined with an array of costimulatory signals
that have different temporal actions and target T cells at different differentiation
stages or activation states. The end result of these multiple costimulation signals
on the T cells response is to regulate the optimal immune response by controlling the initiation, expansion, survival, contraction, and generation of memory.

3. T CELL DEVELOPMENT

In order to understand the importance of costimulation it is appropriate to discuss the development of T cells and the role of costimulation during T cell development. Of course, this means taking into account TCR stimulation and other signals a T cell may encounter, how they differ, and why each signal is required for proper T cell development. Also of importance is the idea that different signaling requirements exist for T cells at each phase of T cell development. The ability of a T cell to signal through its TCR is critical to its survival. An important test a developing T cell faces in the thymus is the ability to begin expressing a functional TCR. Rearrangement of a TCR that is able to recognize and bind self-peptide in the context of self-MHC, with adequate specificity, allows progression of cells into positive selection[7]. Failure to signal through a rearranged TCR leads to death by neglect. Too strong an interaction of the TCR with self-peptide in the context of MHC complex leads to deletion of the thymocyte due to negative selection. However, additional signals other than the TCR are also needed to keep cells alive and allow them to progress down their developmental pathway and mature in the thymus. Several decisions that determine the ultimate phenotype of these potential T cells occur while progressing through the thymus. Choice between the $\alpha\beta$ or $\gamma\delta$ lineage is based on the ability of either the β or γ and δ TCR chains to rearrange, and may require signals other than that of the TCR. Other cell fate decisions as well as proliferation and survival in the thymus may also require additional signals to guide development. The exact influence of costimulation on positive and negative selection is unclear, and little is known in regard to the requirements for costimulation during thymic development. The decision to become a CD4+ or CD8+ T cell seems to be regulated by the ability of CD4 or CD8 to recognize either class II or class I MHC molecules, respectively, during the interaction with TCR, and provide an additional signal through Lck[8,9]. Association of Lck with the cytoplasmic tail of both CD8α and β chains regulates signaling through Lck and development of CD8+ T cells during the selection process in the thymus[10]. Yet the signal through Lck seems to be much more of a requirement for CD4+ T cells as lack of Lck signaling directs cells against the CD4 single positive (SP) T cell phenotype. Also, constitutively active Lck leads to development down the CD4 pathway even in MHC class I restricted TCR transgenic animals[11].

The need for Lck signaling in lineage differentiation, activated through CD4 or CD8 coreceptor, gives one of the first notions that signals other than that of the TCR are required during thymic development. This leads to the question of what other molecules are required during this stage of T cell development and the mechanisms by which they affect T cell activation, proliferation, survival, or death. We will touch on what is known for the role of other forms of costimulation during thymic development, but this field is as of yet wide open as the function of known or novel costimulatory molecules is not clearly defined during thymic development. One molecule that has come to the forefront as critical for

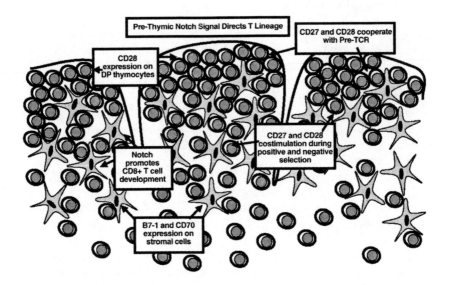

Figure 1. Costimulatory molecules are involved in early stages of T cell development. Signaling through the Notch pathway seems to be one of the earliest signals a T cell progenitor receives. Both CD27 and CD28 are present on cells committed to the T cell lineage even prior to CD3 expression. Several phases in thymic progression may be regulated by costimulatory signals in which specific signals play a proliferative or anti-apoptotic role in T cell development.

thymic development is the Notch family. Notch 1 and 2 receptors are essential for normal embryonic development of several organ systems, including the production of hematopoietic stem cells (HSCs)[12-15]. However, this may be due to the inability of Notch-deficient animals to form proper organs such as bone or liver in which lymphogenesis would occur. Although Notch receptors and their ligands are widely expressed beyond T cell populations, they seem to have specific effects on lymphocyte differentiation. Lymphocytes are particularly affected by Notch, as Notch-transfected HSCs are unable to develop into B lymphocytes and are directed preferentially into the CD8+ T cell lineage[16]. In this case Notch costimulation is required to maintain cells in an undifferentiated state so that they may continue to proliferate and provide progeny cells, which will then differentiate into T precursors[14]. Notch is also critical to the transition between immature double-positive (DP) thymocytes and mature CD4+ or CD8+ SP naive T cells. Notch Delta ligands are thought to synergize with TCR signals and influence this lineage decision as Delta 1-expressing stromal cells can promote maturation of CD8+ cells but not CD4+ T cells[17,18]. Not only does this provide evidence for the necessity of stimulatory signals in addition to TCR signaling in all stages of T cell development, but it also introduces the idea of synergy of the TCR signal with costimulatory signals. Therefore, we must consider that not only the source of the signal but the strength as well is important in initiating and defining the specific response induced.

Here, however, we will discuss more familiar forms of costimulation, as their requirements and functions are not nearly fully understood. Although much attention has been paid to several factors involved in costimulation, we have merely scratched the surface of understanding the entirety of costimulatory mechanisms. We will summarize some of the important aspects of what we do know and discuss important questions yet to be answered. Our interests are rooted in the functions of Ig superfamily member CD28 and TNFR super-family member CD27 during CD8+ T cell responses. These two molecules will be discussed in detail for each phase of CD8+ T cell life cycle starting with thymic development and proceeding through primary activation, memory formation, and antigenic rechallenge, discussing recent findings and implications for future directions. The focus of this review will be on T cells as they progress through different thymocyte development stages and how signals from the CD28 and CD27 molecules in addition to TCR stimulation guide and support this progression.

When T cell costimulation is the topic, one cannot avoid to start the discussion from the classical costimulation molecule CD28. CD28 is the most intensely studied and well-characterized costimulatory molecule of T cells, and yet many aspects of its function are as of yet undefined. This is easily exemplified by the paucity of data related to the function of CD28 in thymic development. The first indication that CD28 may be required for thymic development was its expression on mature human CD3+ thymocytes as well as its ability to be induced on CD3– thymocytes upon PMA stimulation[19]. Addition of anti-CD28 antibodies to CD3+ thymocytes increases thymidine incorporation mediated by CD3 stimulation but has little effect on CD3– thymocytes. This work was furthered by the finding of BB-1 (B7-1), a known CD28 ligand, on thymic stromal cells and the ability of CD28 signaling to induce proliferation of DP thymocytes in the presence of CD3 antibody crosslinked to either CD4 or CD8[20]. These data somewhat conflict with what would be expected in the mouse thymus, as CD28 is expressed most prevalently on immature thymocytes[21]. Proliferation of the SP mature thymocytes is induced only after the decision has been made toward one phenotype or the other. The signal from CD28, however, may play an instructive role in double-negative (DN) TCR positive or DP TCR intermediate populations, where it is highly expressed prior to the proliferative signal transduced in SP mature T cells.

Since thymocytes development requires alternating phases of proliferation, cell survival decisions or differentiation progression, one can reason that receptor signaling and signal transduction during different phases of thymic progression may lead to non-identical results. Each thymocyte population has distinct phenotypic markers, as well as different functional and proliferative capacities. The most important decision in thymic development is that of life/progression through to the next phase or death. Thus the role of CD28 in thymocytes may be dictated by timing throughout T cell development and other signals such as that from the TCR with which it is synergizing. As DN1 (CD44+CD25–) progenitor cells enter the thymus they are committed to the lymphoid lineage, but not yet to that of a T cell. The ability to respond to IL-2 and acquisition of CD25 expression commits these cells to the T cell lineage (DN2). Once cells begin to rearrange the TCR, expression and signaling through CD28 seems to enhance proliferation and survival of DN3 thymocytes; however, this may only be on those cells that are able to properly rearrange a functional pre-TCR, as approximately 20% of thymocytes do not produce functional TCR β chains and undergo apoptosis. However, mice lacking CD28 or B7 have decreased DN2 and DN3 populations with an increase in cells of the DN4 phenotype. Although 80% of DN3 cells express the CD28 counterpart CTLA-4, CTLA-4-deficient animals have otherwise normal thymocytes populations. Both B7- and CD28-deficient DN thymocytes also have increased apoptotic rates and increased TCR expression, which correlate to a block in mitotic progression[22]. Loss of costimulation may correlate to a loss in cell cycle progression; however, the increase in TCR may not be a direct result of upregulation of TCR due to a lack of CD28 costimulation. As previously mentioned, the strength of signal may be the key element in determining the effect of any given stimulus. The selection of cells with higher TCR expression may be the result of a survival advantage of cells capable of increased TCR signaling in the context of absent CD28 signaling and not directly from the lack of signaling through CD28.

Double-transgenic animals for both CD28 and B7-2 exhibit an interesting phenotype in which DN4 populations are also significantly increased[23]. These animals have an overall increase in DN thymocytes and a decrease in the proportion of mature SP thymocytes. Upon closer examination of a seemingly normal population of DP thymocytes, a majority of the DP population does not express TCRβ. This model is also interesting as it rescues thymocytes in the Rag-2- or CD3ε-deficient animals from arrest in the DN phase of thymic development passing the pre-TCR checkpoint, but not as far as to allow for production of SP populations[23]. The importance of CD28 in the progression through positive and negative selection is also of great interest. Animals lacking CD28 or B7 also exhibit higher proportions of SP thymocytes and mature T cells[23,24]. This was demonstrated in both TCR transgenic and wild-type animals, which suggests either an increase in positive selection or a decrease in negative selection. Transduction of a strong enough signal in response to ligation of the TCR on self-MHC/peptide is critical for progression through positive selection. The strength of signal theory, however, would argue against the idea that loss of costimulation would increase positive selection. Also, evidence that a deletion event during negative selection requires costimulation that can be provided by CD28 seems much more plausible. In this case greater cumulative signaling by addition of costimulation can be deleterious as negative selection prevents strong autoreactivity. Still, CD28-B7 interactions play a role in thymic development, although the mechanisms are not clearly defined. Whether CD28 acts merely as an amplification of the TCR or other signals, influences survival and proliferation independently, or has other downstream effects needs to be determined. The importance of timing seems to be critical, and the restricted expression of CD28 and its ligands may also be of relevance.

TNFR superfamily member CD27 has also been identified as an important regulator of T cell development but is not nearly as well described as CD28. CD27 is expressed on most thymocytes, save a population of DN CD25– cells[25]. The expression of CD27 ligand (CD70) has also been demonstrated on thymic stromal cells in addition to activated lymphocytes and dendritic cells (DCs)[26]. Signaling through CD27 helps promote differentiation of DN4 cells in the presence of pre-TCR stimulation. Blocking CD27 in Rag- or TCRα-deficient mice leads to decreased expansion of DP thymocytes. CD27-deficient mice also exhibit in the periphery CD4+ and CD8+ T cells that are unable to proliferate in response to CD3 with or without CD28 stimulation[27]. Thus CD27, like CD28, also plays an important role in thymic development, which is not clearly defined. Again, the nature of the CD27 signal is also undefined in whether it acts independently of or in concert with TCR signals.

4. ANTIGEN-SPECIFIC T CELL RESPONSES

The ability of CD8+ T cells to expand, differentiate, and kill in response to a specific antigenic stimulation is the basis of cell-mediated immunity. The idea that antigenic stimulus alone is not sufficient to extract this ability is both necessary and intriguing. The need for costimulation protects us against autoreactivity, yet hinders our ability to spontaneously defend against insult. Our understanding of the how, why, what, and when of costimulation is gradually developing and helping us understand the best ways to protect ourselves against lethal attack by pathogens. Turning to costimulation as a possible adjuvant for provoking robust immune responses has led us to the pursuit of defining the actions and mechanisms of costimulatory molecules such as CD28 and CD27 during a CD8+ T cell response. Development of new experimental approaches has allowed a better understanding of the function of costimulation, however much is left to be defined.

Recently, much attention has been placed on visualization of the interaction between CD8+ T cells and antigen-presenting cells (APCs) in the context of antigen presentation[28-32]. As mentioned above, such an interaction provides important costimulation signals to the T cells in addition to antigen. These publications have demonstrated the interactions between antigen-specific T cells and APCs in the period after antigenic encounter throughout the priming of T cells in a secondary lymph node. This has helped us to visualize the dynamics of that interaction and establish a better anatomical understanding for the cellular and molecular interactions that occur early in an immune response. Utilizing the technique of two-photon microscopy may lead to a much better understanding of both the TCR–MHC interactions and costimulation that occur during efficient priming of an immune response. However, we must currently depend on indirect evidence from established techniques to explain these interactions. Here we will highlight significant findings and questions that remain unanswered.

CD28 is expressed on all human peripheral blood CD4+ T cells and in about 50% of CD8+ T cells. In mature T cells CD28 signaling can induce the transcription of IL-2, the expression of IL-2Rα chain (CD25) and entry into the cell cycle[4,5,33]. CD28 costimulation has been shown to enhance T cell survival by upregulating the anti-apoptotic member of the Bcl-2 family, Bcl-xL, and c-FLIPshort[34,35]. CD28 signaling also enhances glucose uptake and glycolysis, thus allowing T cells to meet the energetic demands required during an immune response[36]. CD28 was initially shown to increase RNA expression levels of several T cell-derived cytokines in the presence of optimal CD3 stimulation even in the presence of the immunosuppressant cyclosporine A (CSA)[37]. The ability of CD28 to overcome CSA treatment and induce much greater amounts of IL-2 as well as proliferation as compared to CD3, or PMA and calcium ionophore, was taken to mean that CD28 initiated a unique signaling pathway that qualitatively complimented TCR stimulation. CD28 was also shown to increase proliferation and prevent apoptosis in the presence of suboptimal CD3 stimulation[3]. These studies suggest that CD28 has its own signaling cascade, leading to additional transcription factors and responsible for its gene regulation. More recently the strength of signal theory has gained some support[38,39]. This theory makes the argument that CD28 or other costimulatory signals converge upon the TCR pathway to increase the quantitative signal activating gene transcription events, cell-cycle progression, anti-apoptotic factors, and expression of other costimulatory molecules (Figure 2). Analyzing the effect of CD28 on proliferative response and cell division using CFSE labeling provided additional evidence supporting the strength of signal hypothesis. Addition of CD28 to splenocyte cultures increased the sensitivity of T cell proliferation to TCR engagement. However, CD28 costimulation did not change the relative number of cells undergoing cell-cycle progression[40]. Other experiments show increased survival, antigenic response, and proliferative capacity, and decreased doubling time at suboptimal antigen concentrations[41]. Since the signaling cascade specific for CD28 downstream of the most proximal to the receptor signals has not yet been fully identified, and some signaling components of this downstream pathway are shared by those activated by the signaling pathways of the TCR, this theory holds significant weight.

Interestingly, CD28-deficient mice are phenotypically normal with regard to splenic T and B cell populations[42]. Basal immunoglobulin concentrations are decreased in the absence of CD28 with IgG1 and IgG2b isotypes being severely diminished, yet IgG2a being slightly increased[42]. Also, mitogenic stimulation of T cells resulted in decreased proliferation and IL-2 production, which is only partially restored by addition of exogenous IL-2[42]. In the context of antiviral

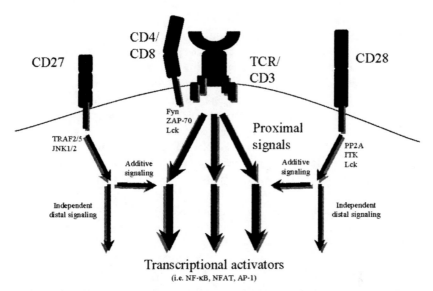

Figure 2. Signals from costimulatory molecules may be additive or independent to those of the TCR. The strength of signal hypothesis dictates that signals from costimulation amplifies the TCR signal and helps to activated the same set of transcription factors. Independent proximal signals have been identified for both CD27 and CD28, but unique transcriptional activators have not been identified. CD27 and CD28 do seem to act independently yet both function synergistically with TCR signals.

responses, CD28 was originally thought to not be required for antiviral CD8+ T cell responses. This was based on studies using CD28-deficient animals infected with lymphocytic choriomeningitis virus (LCMV) and that showed that CD28-deficient animals had normal anti-LCMV CD8+ T cell responses[42]. CD4+ T cell responses to vesicular stomatitis virus (VSV), however, were diminished even though CD8+ T cell cytotoxicity and DTH responses after LCMV infection remained intact. Other studies have shown that CD28 is required for anti-VSV and other virus-specific CD8+ T cell responses, as we discuss below[43-45]. This points to the importance of the infection model in examining the role of costimulation. LCMV infection results in high antigen loads and infection of secondary lymphoid organs that may override the need for costimulation in vivo. This would subscribe to the strength of signal model in that very strong antigenic stimuli through the TCR could replace complimentary signals from other costimulatory molecules. However, the presence of antigen past initiation of a T cell response is not required in some models[46,47]. Therefore, the amount and source of antigen can influence the need for CD28 costimulation and cloud the importance of costimulation in physiological conditions.

The role of CD28 in the context of an antigen-specific response is most relevant when talking about the value of costimulation in adjuvant or therapeutic application. Several models have been used to examine the role of CD28 in CTL responses to viral infections[45,42,48–52]. CD28 is required for CTL responses to influenza, VSV, and low-virulence vaccinia virus, but not LCMV infection. Differences between viral antigenic load/TCR stimulation, viral tropism, localization, and degree of inflammation may account for the differential need for costimulation in these different viral infection models[43]. As mentioned above, the LCMV model differs in both the high antigenic load that is produced as well as the site of infection. LCMV will infect secondary lymphoid organs and cause non-specific inflammation and bystander activation of lymphocytes. Activated bystanders may then promote the inflammatory environment and decrease competition for costimulatory resources and cytokines. In other peripheral tissues where costimulation is scarce, CTL responses may be more dependent upon the presence of CD28 during an ongoing response or priming in the presence of CD28 in the local lymph node. The requirement does not seem to be dependent on the avidity of TCR binding to the peptide/MHC complex as CD8+ T cell responses to the same LCMV-glycoprotein epitope cannot be mounted by CD28-dependent recombinant vaccinia virus that expresses LCMV-glycoprotein, yet such a response can be mounted against CD28-independent LCMV virus. Therefore, the avidity to this peptide/MHC complex is not the critical factor, as LCMV can mount a CD8 response in the absence of CD28 whereas vaccine virus cannot. This may resemble the requirement for CD4+ T cell help, but correlations between the two have not yet been made. It has also been shown that other costimulatory molecules, such as CD27 and 4-1BB, can replace the need for CD28 in CD8+ cell responses[44,53]. This again supports the strength of signal hypothesis and links several sources of synergistic signals to the TCR.

CD27 costimulation also plays an important role in primary CD8+ T cell responses. The initial characterization of CD27 was as a T cell differentiation antigen that facilitated proliferation of mitogen or CD3-activated T cells[54]. T cells in CD27-deficient animals have a deficiency in the ability to respond to CD3 stimulation in vitro or to influenza infection in vivo[27]. Interestingly, CD27-deficient T cells have no deficit in the ability to divide in response to anti-CD3 with or without CD28 costimulation, as CFSE dilution is similar in cultures of purified T cells from wild-type or CD27-deficient animals. However, proliferation assays show a nearly twofold reduction in T cells from CD27-deficient animals as compared to wild-type littermates. Since the defect is only manifested in reduced thymidine incorporation, this could be explained by reduced survival of activated cells. In CD27-deficient animals, differences in influenza virus-specific CD8+ T cell numbers were observed on days 8 and 10 of the primary response, suggesting that the initial expansion phase was not affected, but the later survival of these cells was. Interestingly, the percentage of CD8+ T cells specific for the immunodominant NP(366-374) epitope was unchanged even though the absolute numbers differed. No differences were otherwise observed in the ability of T cells to express IFN-γ or to kill peptide-loaded targets; thus the effector function of the cells that did expand was not affected by lack of CD27[27]. This may reflect a specific subset of cells that require CD27 costimulation or that do not compete as well for other signals.

Two models showed similar results with over-stimulation of CD27[55-57]. CD70 transgenic animals that constitutively express the ligand on B lymphocytes under the control of the CD19 promoter develop nearly normally with regard to T cell populations. In the thymus, however, CD27 is downregulated as well as in the spleen and peripheral lymph nodes, which also contain higher numbers of both CD4+ and CD8+ T cells than wild type. T cells in the periphery appear to be blasting and have an effector/memory phenotype, CD44^{hi}CD62L^{lo}, by 4 weeks of age. Stimulated T cells expressed increased levels of IFN-γ but not TNFα or IL-2. Increased IFN-γ expression gradually led to decreased immunoglobulin levels and eventually depletion of B cell populations, which was abrogated in CD70tg-IFN-γ$^{-/-}$ mice[56]. These mice also showed increased spontaneous proliferation, and cell-cycle progression, of peripheral T cell population although they were less able to proliferate in response to CD3 antibody and were less able to respond to PHA as the mice aged. Aged animals showed a depletion of naive and eventually all CD3+ T cell populations, leading to an inability to mount a primary antigen-specific T cell response and premature death[55]. Recombinant human Fc-CD70 increased antigen-specific proliferation as well as IFN-γ and IL-2 production in the presence or absence of B7.1/B7.2-blocking antibodies in vitro[57]. Injection of Fc-CD70 along with antigen or antigen and LPS dramatically increased the magnitude of both primary and secondary antigen-specific responses by increasing cell division and effector function[57]. CD70 expressed by an unusual population of non-hematopoietic cells in the lamina propria has also been shown to play an important role in protecting against infection by *Listeria monocytogenes*. These antigen-presenting phagocytic cells promoted proliferation and IFN-γ production of antigen-specific CD8+ T cells in gut mucosa that was abrogated by blocking CD70[58]. Thus CD70 promotes CD8+ T cell proliferation at physiological and non-physiological levels and is shown to be important in antigen-specific expansion in response to multiple infections.

CD27 signaling was also determined to be important for CD8+ T cell responses in the absence of CD28. As shown in similar experiments with 4-1BB[44,59], CD27 stimulation was shown to overcome blocking of B7 molecules to induce T cell proliferation[57]. Blocking CD70 prolonged cardiac allograft survival, decreased CD8+ T cell numbers and IFN-γ production in response to the graft, and enhanced graft survival in CD28-deficient mice[53]. CD27-deficient mice also accumulated fewer influenza-specific CD8+ T cells in the lung as compared to CD28-deficient mice. Whereas CD28 seems to be more important in lymphoid organs, CD27 seems to enhance survival of antigen-activated CD8+ T cells in non-lymphoid tissues. CD27-deficient animals had decreased numbers of cells accumulating in the lung without proliferation of these cells being affected[52]. This identifies CD27 as a CD28-independent survival factor for activated CD8+ T cells.

The lack of studies on the requirement of CD27 costimulation in other viral infection models leaves unanswered the question of whether some infections can also be independent of CD27 costimulation, similar to what occurs with CD28 costimulation and LCMV infection. CD27 costimulation does seem to be less important in the spleen as compared to CD28, and yet there does not seem to be an additive effect when both CD28 and CD27 are absent[52]. Thus differences and similarities seem to exist between CD27 and CD28 costimulation during viral infections which, however, have yet to be fully elucidated.

5. MEMORY, ANTIGENIC RECHALLENGE, AND SECONDARY RESPONSES

The ability to develop protective memory is the ultimate goal of vaccination or stimulation of the adaptive immune system. Soluble factors such as cytokines have been traditionally thought of as the survival stimuli of the immune system. Since the characterization of IL-2 as a T cell growth factor, this field has expanded and grown to include an immense number of factors whose functions and targets vary. Recently several cytokines have been given special attention for their role in keeping T cells alive. In particular, IL-7, IL-15, in addition to IL-2 are known to have related, but distinct, roles in CD8+ T cell survival and in memory generation and maintenance following viral infections[60–71]. However, the role of CD28 and CD27 costimulation in the production of CD8+ T cell memory has been all but ignored. The lack of data is somewhat due to the models that are available to study these molecules. Knockout animals for either the CD28 or CD27 costimulatory molecule have proven extremely useful in our understanding of their function during primary CD8+ T cell responses. Data discussed in §§3 and 4 of this review, however, clearly demonstrate that the usefulness of those models in elucidating the role of costimulation plays in memory development is compromised by the fact that primary responses are affected. Given that memory CD8+ T cell responses are influenced and determined to some extent by the magnitude of the primary response[72–75] and that these primary responses are diminished by the lack of CD28 or CD27 in knockout animals[27,42,52], it is clear that it is very difficult, if not impossible, to dissect any direct effect of CD28 or CD27 costimulation on memory generation or maintenance or quality of response of memory upon rechallenge in these knockout animals. Furthermore, even though the development of memory may not require primary effector cell expansion[46,47,76,77], intrinsic deficits or defects in thymic development of T cells that occur in CD28 and CD27 knockout animals[19,22,24,25,27,54,78–81] may be manifested as defects in peripheral mature T cells that result in impaired memory responses. The addition of cloned exogenous cytokine, agonist, or antagonist monoclonal antibodies directed against costimulatory receptors or ligands or inhibitory fusion proteins is a another approach that can be taken to study the role of CD28 or CD27 costimulation in memory responses. Using such reagents one can study the memory CD8+ T cell response in animals that have undergone an intact primary response.

Upon contraction of a primary immune response, the vast majority of effector T cells undergo apoptosis, leaving behind a population of cells representing memory[46,72,82-86]. Much work has been done to elucidate the mechanisms involved in the production and maintenance of these cells. A theory of linear differentiation from naive to effector and finally memory[87,88] may be giving way to one where a defining event that occurs early in antigen encounter results in memory production bypassing effector differentiation[72,76,85,89-91]. However, costimulatory requirements for the production and maintenance of memory populations have yet to be fully determined. Recent examination of CD4+ T cell help has determined that, although CD4+ T cells may not be required for primary CTL expansion in some infection models, they are required in maintaining memory CD8+ T cells[92-98]. This may be through direct CD4–CD8 T cell interactions via CD40–CD154 or through CD4 help in licensing APCs. Other factors such as cytokines and surface-bound receptors may also play a role that has not yet been identified. Additionally, though the primary CD8+ T cell response may occur after infection has been cleared, the requirements of CD8+ T cell interactions nonspecifically or with antigen-loaded APCs have not been ruled out during expansion nor memory maintenance. A useful transgenic model (DTR-CD11c transgenic animals) has been developed in which CD11c+ dendritic cells can be selectively depleted throughout an immune response[99]. Although this may not directly identify the factor or factors that mediate memory production, determining the source of those factors will be of value. Indeed, using the DTR-CD11c transgenic animals it was recently demonstrated that DCs are required during the secondary response for the optimal antiviral CD8+ T cell response against influenza virus rechallenge[100]. The implications of these findings are that costimulation or cytokines derived from DC are important for the reactivation, expansion, or survival of memory CD8+ T cells.

CD28 does not seem to be required to maintain memory populations in LCMV-infected animals[101]. Although CD8+ T cells are decreased during the primary response and memory is reduced twofold in CD28-deficient animals, activation threshold, homeostatic proliferation, and protective immunity to lethal challenge were maintained[101]. Influenza infections, which require CD28 costimulation, produce less memory in CD28-deficient or CTLA-4-Ig transgenic animals[50,59]. However, these two models differ in the requirement for CD28 during the secondary response and reactivation of memory CD8+ T cells. Replacement of CD28 with 4-1BB stimulation recovers primary CD8+ T cell responses in CD28-deficient animals[44,59]. In CD28-deficient animals 4-1BB stimulation is sufficient to restore secondary responses even though memory cells still lack CD28 during reactivation. However, blocking B7 during restimulation of primed

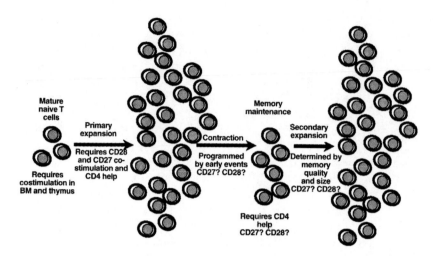

Figure 3. Requirements for costimulation and CD4 help during CD8+ T cell development and differentiation. Some aspects of CD8+ T cell requirements have been defined for different stages of the immune response and T cell development. Known requirements for CD4+ T cells help during a CD8+ T cell response may parallel costimulatory requirements. This relationship needs to be defined as costimulation may directly affect CD8+ T cells or affect them indirectly through APCs or CD4+ T cells. Clarifying the role of CD27 and CD28 may open new windows of therapeutic and adjuvant usage of costimulatory molecules.

splenocytes decreases expansion[50]. *Listeria monocytogenes* infection of CD28-deficient mice results in decreased CD8+ T cell number and impaired effector function[102]. Secondary responses in these animals reflected a decrease in the primary response, but the remaining cells seem to have normal effector function, meaning the secondary expansion was not specifically inhibited by the lack of CD28. From the above it is apparent that the role of CD28 during memory CD8+ T cell reactivation and secondary responses has yet to be fully elucidated.

Currently, there are very few studies that have examined CD8+ T cell requirement for CD27 costimulation. CD27-deficient animals have a defective primary response, resulting in less memory and decreased secondary expansion[27,52,103]. CD27, 4-1BB, and OX40 each play a significant role in the primary CD8+ T cell response to influenza virus. In affecting the primary response, memory formation was also impaired as well as recall responses. Hendriks et al. went a step further in examining the quality of memory expansion. Transfer of equivalent numbers of memory CD8+ T cells from CD27-, 4-1BBL-, or OX40L-deficient animals demonstrated the inability of these cells to respond equivalently to reinfection. However, a lack of CD27 signaling during the secondary response was not ruled out[103]. Even if primary expansion does not define the size of memory populations, CD27 may be required during the maintenance phase as opposed to the effector phase. CD27 costimulation may parallel that of

CD4+ T cell help and be required for maintenance of antigen-specific memory CD8+ T cells[98]. However, stimulation with soluble CD70 enhances secondary responses to OVA antigen[57]. This does not necessarily define the role of CD27 in memory or secondary responses, as excessive costimulation may enhance a response even if it is not physiologically required. Monoclonal antibody treatments of normally primed animals may clarify the role of CD27 in the secondary response.

Again, defining the role of costimulation in memory production and maintenance is expected to be complex. New models and innovative manipulation of existing models will be needed. Knockout models may prove useful for transfer of normally primed cells, but caveats to those experiments must be taken into account when analyzing the results.

6. SUMMARY AND CONCLUSIONS

Although the requirement of CD28 and CD27 costimulation has been clearly demonstrated during primary CD8+ T cell responses and this costimulation acts by providing proliferation and survival cues to naive CD8+ T cells, a number of questions also arise from these studies. Is the requirement for CD28 and CD27 costimulation restricted to the initiation of the immune response in the lymph nodes, where presumably the initial contact between naive CD8+ T cell and DC occurs? What is the purpose of the dramatic influx of DC to sites of inflammation such as the lung during influenza virus infection and the formation of inflammatory BALT (iBALT)?[104] Are such DC at the site of inflammation and at later stages of the immune response providing cytokines or costimulation to effector CD8+ T cells? If DC are required for optimal secondary responses[100], is CD28 costimulation the missing signal or is it other members of the B7:CD28 family or TNF family? Given that a number of investigators are actively addressing these questions, the answers we expect will be soon to come and open exciting new opportunities for immune enhancement or dampening strategies and vaccine adjuvants.

7. REFERENCES

1. P. Bretscher and M. Cohn. A theory of self-nonself discrimination. *Science* **169**(950):1042–1049 (1970).
2. D.L. Mueller, M.K. Jenkins and R.H. Schwartz. Clonal expansion versus functional clonal inactivation: a costimulatory signalling pathway determines the outcome of T cell antigen receptor occupancy. *Annu Rev Immunol* **7**:445–480 (1989).
3. F.A. Harding, J.G. McArthur, J.A. Gross, D.H. Raulet and J.P. Allison. CD28-mediated signalling co-stimulates murine T cells and prevents induction of anergy in T-cell clones. *Nature* **356**(6370):607–609 (1992).
4. R.J. Greenwald, G.J. Freeman and A.H. Sharpe. The B7 family revisited. *Annu Rev Immunol* **23**:515–548 (2005).

5. J.L. Riley and C.H. June. The CD28 family: a T-cell rheostat for therapeutic control of T-cell activation. *Blood* **105**(1):13–21 (2005).

6. M. Croft. Co-stimulatory members of the TNFR family: keys to effective T-cell immunity? *Nat Rev Immunol* **3**(8):609–620 (2003).

7. C.A. Janeway Jr. A tale of two T cells. *Immunity* **8**(4):391–394 (1998).

8. K. Laky and B.J. Fowlkes. Receptor signals and nuclear events in CD4 and CD8 T cell lineage commitment. *Curr Opin Immunol* **17**(2):116–121 (2005).

9. R.N. Germain. T-cell development and the CD4-CD8 lineage decision. *Nat Rev Immunol* **2**(5):309–322 (2002).

10. H.Y. Irie, M.S. Mong, A. Itano, M.E. Crooks, D.R. Littman, S.J. Burakoff and E. Robey. The cytoplasmic domain of CD8 beta regulates Lck kinase activation and CD8 T cell development. *J Immunol* **161**(1):183–191 (1998).

11. G. Hernandez-Hoyos, S.J. Sohn, E.V. Rothenberg and J. Alberola-Ila. Lck Activity Controls CD4/CD8 T Cell Lineage Commitment. *Immunity* **12**(3):313 (2000).

12. I. Maillard, T. Fang and W.S. Pear. Regulation of lymphoid development, differentiation, and function by the Notch pathway. *Annu Rev Immunol* **23**:945–974 (2005).

13. H. von Boehmer. Notch in lymphopoiesis and T cell polarization. *Nat Immunol* **6**(7):641–642 (2005).

14. A. Sambandam, I. Maillard, V.P. Zediak, L. Xu, R.M. Gerstein, J.C. Aster, W.S. Pear and A. Bhandoola. Notch signaling controls the generation and differentiation of early T lineage progenitors. *Nat Immunol* **6**(7):663–670 (2005).

15. E.A. Robey and J.A. Bluestone. Notch signaling in lymphocyte development and function. *Curr Opin Immunol* **16**(3):360–366 (2004).

16. B. Varnum-Finney, L. Xu, C. Brashem-Stein, C. Nourigat, D. Flowers, S. Bakkour, W.S. Pear and I.D. Bernstein. Pluripotent, cytokine-dependent, hematopoietic stem cells are immortalized by constitutive Notch1 signaling. *Nat Med* **6**(11):1278–1281 (2000).

17. D.J. Izon, J.A. Punt, L. Xu, F.G. Karnell, D. Allman, P.S. Myung, N.J. Boerth, J.C. Pui, G.A. Koretzky and W.S. Pear. Notch1 regulates maturation of CD4+ and CD8+ thymocytes by modulating TCR signal strength. *Immunity* **14**(3):253–264 (2001).

18. A. de La Coste, E. Six, N. Fazilleau, L. Mascarell, N. Legrand, M.P. Mailhe, A. Cumano, Y. Laabi and A.A. Freitas. In vivo and in absence of a thymus, the enforced expression of the Notch ligands delta-1 or delta-4 promotes T cell development with specific unique effects. *J Immunol* **174**(5):2730–2737 (2005).

19. L.A. Turka, J.A. Ledbetter, K. Lee, C.H. June and C.B. Thompson. CD28 is an inducible T cell surface antigen that transduces a proliferative signal in CD3+ mature thymocytes. *J Immunol* **144**(5):1646–1653 (1990).

20. L.A. Turka, P.S. Linsley, R. d. Paine, G.L. Schieven, G.B. Thompson and J.A. Ledbetter. Signal transduction via CD4, CD8, and CD28 in mature and immature thymocytes: implications for thymic selection. *J Immunol* **146**(5):1428–1436 (1991).

21. J.A. Gross, E. Callas and J.P. Allison. Identification and distribution of the costimulatory receptor CD28 in the mouse. *J Immunol* **149**(2):380–388 (1992).

22. X. Zheng, J.-X. Gao, X. Chang, Y. Wang, Y. Liu, J. Wen, H. Zhang, J. Zhang, Y. Liu and P. Zheng. B7-CD28 interaction promotes proliferation and survival but suppresses differentiation of CD4-CD8- T cells in the thymus. *J Immunol* **173**(4):2253–2261 (2004).

23. J.A. Williams, K.S. Hathcock, D. Klug, Y. Harada, B. Choudhury, J.P. Allison, R. Abe and R.J. Hodes. Regulated costimulation in the thymus is critical for T cell de-

velopment: dysregulated CD28 costimulation can bypass the pre-TCR checkpoint. *J Immunol* **175**(7):4199–4207 (2005).

24. M.S. Vacchio, J.A. Williams and R.J. Hodes. A novel role for CD28 in thymic selection: elimination of CD28/B7 interactions increases positive selection. *Eur J Immunol* **35**(2):418–427 (2005).

25. L. Gravestein, W. van Ewijk, F. Ossendorp and J. Borst. CD27 cooperates with the pre-T cell receptor in the regulation of murine T cell development. *J Exp Med* **184**(2):675–685 (1996).

26. K. Tesselaar, Y. Xiao, R. Arens, G.M.W. van Schijndel, D.H. Schuurhuis, R.E. Mebius, J. Borst and R.A.W. van Lier. Expression of the murine CD27 ligand CD70 in vitro and in vivo. *J Immunol* **170**(1):33–40 (2003).

27. J. Hendriks, L.A. Gravestein, K. Tesselaar, R.A. van Lier, T.N. Schumacher and J. Borst. CD27 is required for generation and long-term maintenance of T cell immunity. *Nat Immunol* **1**(5):433–440 (2000).

28. S. Stoll, J. Delon, T.M. Brotz and R.N. Germain. Dynamic imaging of T cell-dendritic cell interactions in lymph nodes. *Science* **296**(5574):1873–1876 (2002).

29. T.R. Mempel, S.E. Henrickson and U.H. Von Andrian. T-cell priming by dendritic cells in lymph nodes occurs in three distinct phases. *Nature* **427**(6970):154–159 (2004).

30. R.L. Lindquist, G. Shakhar, D. Dudziak, H. Wardemann, T. Eisenreich, M.L. Dustin and M.C. Nussenzweig. Visualizing dendritic cell networks in vivo. *Nat Immunol* **5**(12):1243–1250 (2004).

31. S. Hugues, L. Fetler, L. Bonifaz, J. Helft, F. Amblard and S. Amigorena. Distinct T cell dynamics in lymph nodes during the induction of tolerance and immunity. *Nat Immunol* **5**(12):1235–1242 (2004).

32. A. Lanzavecchia and F. Sallusto. Lead and follow: the dance of the dendritic cell and T cell. *Nat Immunol* **5**(12):1201–1202 (2004).

33. C.H. June, J.A. Ledbetter, M.M. Gillespie, T. Lindsten and C.B. Thompson. T-cell proliferation involving the CD28 pathway is associated with cyclosporine-resistant interleukin 2 gene expression. *Mol Cell Biol* **7**(12):4472–4481 (1987).

34. A.I. Sperling, J.A. Auger, B.D. Ehst, I.C. Rulifson, C.B. Thompson and J.A. Bluestone. CD28/B7 interactions deliver a unique signal to naive T cells that regulates cell survival but not early proliferation. *J Immunol* **157**(9):3909–3917 (1996).

35. S. Kirchhoff, W.W. Muller, M. Li-Weber and P.H. Krammer. Up-regulation of c-FLIPshort and reduction of activation-induced cell death in CD28-costimulated human T cells. *Eur J Immunol* **30**(10):2765–2774 (2000).

36. K.A. Frauwirth, J.L. Riley, M.H. Harris, R.V. Parry, J.C. Rathmell, D.R. Plas, R.L. Elstrom, C.H. June and C.B. Thompson. The CD28 signaling pathway regulates glucose metabolism. *Immunity* **16**(6):769–777 (2002).

37. C.B. Thompson, T. Lindsten, J.A. Ledbetter, S.L. Kunkel, H.A. Young, S.G. Emerson, J.M. Leiden and C.H. June. CD28 activation pathway regulates the production of multiple T-cell-derived lymphokines/cytokines. *PNAS* **86**(4):1333–1337 (1989).

38. O. Acuto and F. Michel. CD28-mediated co-stimulation: a quantitative support for TCR signalling. *Nat Rev Immunol* **3**(12):939–951 (2003).

39. A.V. Gett, F. Sallusto, A. Lanzavecchia and J. Geginat. T cell fitness determined by signal strength. *Nat Immunol* **4**(4):355–360 (2003).

40. A.D. Wells, H. Gudmundsdottir and L.A. Turka. Following the fate of individual T cells throughout activation and clonal expansion: signals from T cell receptor and

CD28 differentially regulate the induction and duration of a proliferative response. *J Clin Invest* **100**(12):3173–3183 (1997).

41. H. Gudmundsdottir, A.D. Wells and L.A. Turka. Dynamics and requirements of T Cell clonal expansion in vivo at the single-cell level: effector function is linked to proliferative capacity. *J Immunol* **162**(9):5212–5223 (1999).

42. A. Shahinian, K. Pfeffer, K.P. Lee, T.M. Kundig, K. Kishihara, A. Wakeham, K. Kawai, P.S. Ohashi, C.B. Thompson and T.W. Mak. Differential T cell costimulatory requirements in CD28-deficient mice. *Science* **261**(5121):609–612 (1993).

43. T.M. Kundig, A. Shahinian, K. Kawai, H.W. Mittrucker, E. Sebzda, M.F. Bachmann, T.W. Mak and P.S. Ohashi. Duration of TCR stimulation determines costimulatory requirement of T cells. *Immunity* **5**(1):41–52 (1996).

44. E.S. Halstead, Y.M. Mueller, J.D. Altman and P.D. Katsikis. In vivo stimulation of CD137 broadens primary antiviral CD8(+) T cell responses. *Nat Immunol* **3**(6):536–541 (2002).

45. S.O. Andreasen, J.E. Christensen, O. Marker and A.R. Thomsen. Role of CD40 ligand and CD28 in induction and maintenance of antiviral CD8+ effector T cell responses. *J Immunol* **164**(7):3689–3697 (2000).

46. V.P. Badovinac, B.B. Porter and J.T. Harty. CD8+ T cell contraction is controlled by early inflammation. *Nat Immunol* **5**(8):809 (2004).

47. G.A. Corbin and J.T. Harty. Duration of infection and antigen display have minimal influence on the kinetics of the CD4+ T cell response to *Listeria monocytogenes* infection. *J Immunol* **173**(9):5679–5687 (2004).

48. C. Zimmermann, P. Seiler, P. Lane and R.M. Zinkernagel. Antiviral immune responses in CTLA4 transgenic mice. *J Virol* **71**(3):1802–1807 (1997).

49. M. Kopf, A.J. Coyle, N. Schmitz, M. Barner, A. Oxenius, A. Gallimore, J.C. Gutierrez-Ramos and M.F. Bachmann. Inducible costimulator protein (ICOS) controls T helper cell subset polarization after virus and parasite infection. *J Exp Med* **192**(1):53–61 (2000).

50. Y. Liu, R.H. Wenger, M. Zhao and P.J. Nielsen. Distinct costimulatory molecules are required for the induction of effector and memory cytotoxic T lymphocytes. *J Exp Med* **185**(2):251–262 (1997).

51. J.K. Whitmire and R. Ahmed. Costimulation in antiviral immunity: differential requirements for CD4(+) and CD8(+) T cell responses. *Curr Opin Immunol* **12**(4):448–455 (2000).

52. J. Hendriks, Y. Xiao and J. Borst. CD27 promotes survival of activated T cells and complements CD28 in generation and establishment of the effector T cell pool. *J Exp Med* **198**(9):1369–1380 (2003).

53. A. Yamada, A.D. Salama, M. Sho, N. Najafian, T. Ito, J.P. Forman, R. Kewalramani, S. Sandner, H. Harada, M.R. Clarkson, D.A. Mandelbrot, A.H. Sharpe, H. Oshima, H. Yagita, G. Chalasani, F.G. Lakkis, H. Auchincloss Jr. and M.H. Sayegh. CD70 signaling is critical for CD28-independent CD8+ T cell-mediated alloimmune responses in vivo. *J Immunol* **174**(3):1357–1364 (2005).

54. R. van Lier, J. Borst, T. Vroom, H. Klein, P. Van Mourik, W. Zeijlemaker and C. Melief. Tissue distribution and biochemical and functional properties of Tp55 (CD27), a novel T cell differentiation antigen. *J Immunol* **139**(5):1589–1596 (1987).

55. K. Tesselaar, R. Arens, G.M. van Schijndel, P.A. Baars, M.A. van der Valk, J. Borst, M.H. van Oers and R.A. van Lier. Lethal T cell immunodeficiency induced

by chronic costimulation via CD27-CD70 interactions. *Nat Immunol* **4**(1):49–54 (2003).

56. R. Arens, K. Tesselaar, P.A. Baars, G.M. van Schijndel, J. Hendriks, S.T. Pals, P. Krimpenfort, J. Borst, M.H. van Oers and R.A. van Lier. Constitutive CD27/CD70 interaction induces expansion of effector-type T cells and results in IFNgamma-mediated B cell depletion. *Immunity* **15**(5):801–812 (2001).

57. T.F. Rowley and A. Al-Shamkhani. Stimulation by soluble CD70 promotes strong primary and secondary CD8+ cytotoxic T cell responses in vivo. *J Immunol* **172**(10):6039–6046 (2004).

58. A. Laouar, V. Haridas, D. Vargas, X. Zhinan, D. Chaplin, R.A. van Lier and N. Manjunath. CD70+ antigen-presenting cells control the proliferation and differentiation of T cells in the intestinal mucosa. *Nat Immunol* **6**(7):698–706 (2005).

59. E.M. Bertram, W. Dawicki, B. Sedgmen, J.L. Bramson, D.H. Lynch and T.H. Watts. A switch in costimulation from CD28 to 4-1BB during primary versus secondary CD8 T cell response to influenza in vivo. *J Immunol* **172**(2):981–988 (2004).

60. C.C. Ku, M. Murakami, A. Sakamoto, J. Kappler and P. Marrack. Control of homeostasis of CD8+ memory T cells by opposing cytokines. *Science* **288**(5466):675–678 (2000).

61. X.C. Li, G. Demirci, S. Ferrari-Lacraz, C. Groves, A. Coyle, T.R. Malek and T.B. Strom. IL-15 and IL-2: a matter of life and death for T cells in vivo. *Nat Med* **7**(1):114–118 (2001).

62. J.T. Tan, B. Ernst, W.C. Kieper, E. LeRoy, J. Sprent and C.D. Surh. Interleukin (IL)-15 and IL-7 jointly regulate homeostatic proliferation of memory phenotype CD8(+) cells but are not required for memory phenotype CD4(+) cells. *J Exp Med* **195**(12):1523–1532 (2002).

63. W.C. Kieper, J.T. Tan, B. Bondi-Boyd, L. Gapin, J. Sprent, R. Ceredig and C.D. Surh. Overexpression of interleukin (IL)-7 Leads to IL-15-independent generation of memory phenotype CD8(+) T cells. *J Exp Med* **195**(12):1533–1539 (2002).

64. T.C. Becker, E.J. Wherry, D. Boone, K. Murali-Krishna, R. Antia, A. Ma and R. Ahmed. Interleukin 15 is required for proliferative renewal of virus-specific memory CD8 T cells. *J Exp Med* **195**(12):1541–1548 (2002).

65. A.D. Judge, X. Zhang, H. Fujii, C.D. Surh and J. Sprent. Interleukin 15 controls both proliferation and survival of a subset of memory-phenotype CD8(+) T cells. *J Exp Med* **196**(7):935–946 (2002).

66. K.S. Schluns and L. Lefrancois. Cytokine control of memory T-cell development and survival. *Nat Rev Immunol* **3**(4):269–279 (2003).

67. Y.M. Mueller, V. Makar, P.M. Bojczuk, J. Witek and P.D. Katsikis. IL-15 enhances the function and inhibits CD95/Fas-induced apoptosis of human CD4+ and CD8+ effector-memory T cells. *Int Immunol* **15**(1):49–58 (2003).

68. K.S. Schluns, W.C. Kieper, S.C. Jameson and L. Lefrancois. Interleukin-7 mediates the homeostasis of naive and memory CD8 T cells in vivo. *Nat Immunol* **1**(5):426–432 (2000).

69. S. Jaleco, L. Swainson, V. Dardalhon, M. Burjanadze, S. Kinet and N. Taylor. Homeostasis of naive and memory CD4+ T cells: IL-2 and IL-7 differentially regulate the balance between proliferation and Fas-mediated apoptosis. *J Immunol* **171**(1):61–68 (2003).

70. E. Maraskovsky, M. Teepe, P. Morrissey, S. Braddy, R. Miller, D. Lynch and J. Peschon. Impaired survival and proliferation in IL-7 receptor-deficient peripheral T cells. *J Immunol* **157**(12):5315–5323 (1996).

71. H. Dooms, E. Kahn, B. Knoechel and A.K. Abbas. IL-2 induces a competitive survival advantage in T lymphocytes. *J Immunol* **172**(10):5973–5979 (2004).

72. S.M. Kaech and R. Ahmed. Memory CD8+ T cell differentiation: initial antigen encounter triggers a developmental program in naive cells. *Nat Immunol* **2**(5):415–422 (2001).

73. M.J. Bevan and P.J. Fink. The CD8 response on autopilot. *Nat Immunol* **2**(5):381–382 (2001).

74. K. Murali-Krishna, J.D. Altman, M. Suresh, D.J. Sourdive, A.J. Zajac, J.D. Miller, J. Slansky and R. Ahmed. Counting antigen-specific CD8 T cells: a reevaluation of bystander activation during viral infection. *Immunity* **8**(2):177–187 (1998).

75. S. Hou, L. Hyland, K.W. Ryan, A. Portner and P.C. Doherty. Virus-specific CD8+ T-cell memory determined by clonal burst size. *Nature* **369**(6482):652–654 (1994).

76. N. Manjunath, P. Shankar, J. Wan, W. Weninger, M.A. Crowley, K. Hieshima, T.A. Springer, X. Fan, H. Shen, J. Lieberman and U.H. von Andrian. Effector differentiation is not prerequisite for generation of memory cytotoxic T lymphocytes. *J Clin Invest* **108**(6):871–878 (2001).

77. D.L. Woodland and M.A. Blackman. Vaccine development: baring the 'dirty little secret'. *Nat Med* **11**(7):715 (2005).

78. A. Wiesmann, R.L. Phillips, M. Mojica, L.J. Pierce, A.E. Searles, G.J. Spangrude and I. Lemischka. Expression of CD27 on murine hematopoietic stem and progenitor cells. *Immunity* **12**(2):193–199 (2000).

79. K. Tesselaar, L. Gravestein, G. van Schijndel, J. Borst and R. van Lier. Characterization of murine CD70, the ligand of the TNF receptor family member CD27. *J Immunol* **159**(10):4959–4965 (1997).

80. R. de Jong, W. Loenen, M. Brouwer, L. van Emmerik, E. de Vries, J. Borst and R. van Lier. Regulation of expression of CD27, a T cell-specific member of a novel family of membrane receptors. *J Immunol* **146**(8):2488–2494 (1991).

81. M. Bowman, M. Crimmins, J. Yetz-Aldape, R. Kriz, K. Kelleher and S. Herrmann. The cloning of CD70 and its identification as the ligand for CD27. *J Immunol* **152**(4):1756–1761 (1994).

82. L.E. Gamadia, E.M. M. van Leeuwen, E.B.M. Remmerswaal, S.-L. Yong, S. Surachno, P.M.E. Wertheim-van Dillen, I.J.M. ten Berge and R.A.W. van Lier. The size and phenotype of virus-specific T cell populations is determined by repetitive antigenic stimulation and environmental cytokines. *J Immunol* **172**(10):6107–6114 (2004).

83. B. Jamieson and R. Ahmed. T cell memory. Long-term persistence of virus-specific cytotoxic T cells. *J Exp Med* **169**(6):1993–2005 (1989).

84. V. Appay, P.R. Dunbar, M. Callan, P. Klenerman, G.M. Gillespie, L. Papagno, G.S. Ogg, A. King, F. Lechner, C.A. Spina, S. Little, D.V. Havlir, D.D. Richman, N. Gruener, G. Pape, A. Waters, P. Easterbrook, M. Salio, V. Cerundolo, A.J. McMichael and S.L. Rowland-Jones. Memory CD8+ T cells vary in differentiation phenotype in different persistent virus infections. *Nat Med* **8**(4):379–385 (2002).

85. F. Sallusto and A. Lanzavecchia. Exploring pathways for memory T cell generation. *J Clin Invest* **108**(6):805–806 (2001).

86. M.J. van Stipdonk, E.E. Lemmens and S.P. Schoenberger. Naive CTLs require a single brief period of antigenic stimulation for clonal expansion and differentiation. *Nat Immunol* **2**(5):423–429 (2001).

87. A.L. Marzo, K.D. Klonowski, A. Le Bon, P. Borrow, D.F. Tough and L. Lefrancois. Initial T cell frequency dictates memory CD8+ T cell lineage commitment. *Nat Immunol* **6**(8):793–799 (2005).

88. M.A. Williams and M.J. Bevan. T cell memory: fixed or flexible? *Nat Immunol* **6**(8):752 (2005).

89. V.P. Badovinac and J.T. Harty. Memory lanes. *Nat Immunol* **4**(3):212–213 (2003).

90. E.J. Wherry, V. Teichgraber, T.C. Becker, D. Masopust, S.M. Kaech, R. Antia, U.H. von Andrian and R. Ahmed. Lineage relationship and protective immunity of memory CD8 T cell subsets. *Nat Immunol* **4**(3):225–234 (2003).

91. S.M. Kaech, J.T. Tan, E.J. Wherry, B.T. Konieczny, C.D. Surh and R. Ahmed. Selective expression of the interleukin 7 receptor identifies effector CD8 T cells that give rise to long-lived memory cells. *Nat Immunol* **4**(12):1191–1198 (2003).

92. C.M. Smith, N.S. Wilson, J. Waithman, J.A. Villadangos, F.R. Carbone, W.R. Heath and G.T. Belz. Cognate CD4(+) T cell licensing of dendritic cells in CD8(+) T cell immunity. *Nat Immunol* **5**(11):1143–1148 (2004).

93. G.T. Belz, H. Liu, S. Andreansky, P.C. Doherty and P.G. Stevenson. Absence of a functional defect in CD8+ T cells during primary murine gammaherpesvirus-68 infection of I-Ab$^{-/-}$ mice. *J Gen Virol* **84**(2):337–341 (2003).

94. M.J. Bevan. Helping the CD8(+) T-cell response. *Nat Rev Immunol* **4**(8):595–602 (2004).

95. A.L. Marzo, V. Vezys, K.D. Klonowski, S.-J. Lee, G. Muralimohan, M. Moore, D.F. Tough and L. Lefrancois. Fully functional memory CD8 T cells in the absence of CD4 T cells. *J Immunol* **173**(2):969–975 (2004).

96. D.J. Shedlock and H. Shen. Requirement for CD4 T cell help in generating functional CD8 T cell memory. *Science* **300**(5617):337–339 (2003).

97. J.C. Sun and M.J. Bevan. Defective CD8 T cell memory following acute infection without CD4 T cell help. *Science* **300**(5617):339–342 (2003).

98. J.C. Sun, M.A. Williams and M.J. Bevan. CD4+ T cells are required for the maintenance, not programming, of memory CD8+ T cells after acute infection. *Nat Immunol* **5**(9):927–933 (2004).

99. S. Jung, D. Unutmaz, P. Wong, G. Sano, K. De los Santos, T. Sparwasser, S. Wu, S. Vuthoori, K. Ko, F. Zavala, E.G. Pamer, D.R. Littman and R.A. Lang. In vivo depletion of CD11c(+) dendritic cells abrogates priming of CD8(+) T cells by exogenous cell-associated antigens. *Immunity* **17**(2):211–220 (2002).

100. D.J. Zammit, L.S. Cauley, Q.M. Pham and L. Lefrancois. Dendritic cells maximize the memory CD8 T cell response to infection. *Immunity* **22**(5):561–570 (2005).

101. M. Suresh, J.K. Whitmire, L.E. Harrington, C.P. Larsen, T.C. Pearson, J.D. Altman and R. Ahmed. Role of CD28-B7 interactions in generation and maintenance of CD8 T cell memory. *J Immunol* **167**(10):5565–5573 (2001).

102. H.-W. Mittrucker, M. Kursar, A. Kohler, R. Hurwitz and S.H.E. Kaufmann. Role of CD28 for the generation and expansion of antigen-specific CD8+ T lymphocytes during infection with *Listeria monocytogenes*. *J Immunol* **167**(10):5620–5627 (2001).

103. J. Hendriks, Y. Xiao, J.W.A. Rossen, K.F. van der Sluijs, K. Sugamura, N. Ishii and J. Borst. During viral infection of the respiratory tract, CD27, 4-1BB, and OX40 col-

lectively determine formation of CD8+ memory T cells and their capacity for secondary expansion. *J Immunol* **175**(3):1665–1676 (2005).

104. J.E. Moyron-Quiroz, J. Rangel-Moreno, K. Kusser, L. Hartson, F. Sprague, S. Goodrich, D.L. Woodland, F.E. Lund and T.D. Randall. Role of inducible bronchus associated lymphoid tissue (iBALT) in respiratory immunity. *Nat Med* **10**(9):927–934 (2004).

12

CD38: AN ECTO-ENZYME AT THE CROSSROADS OF INNATE AND ADAPTIVE IMMUNE RESPONSES

Santiago Partidá-Sánchez, Laura Rivero-Nava,
Guixiu Shi, and Frances E. Lund

1. INTRODUCTION

No one would dispute that intracellular enzymes such as kinases and phosphatases play critical roles in regulating the development, activation, differentiation, and survival of lymphocytes[1]. However, it is less well appreciated that cells of the immune system also express many membrane-associated ecto-enzymes that have the potential to regulate immune cell function. Ecto-enzymes have their active sites located on the outside of the cell and therefore must utilize substrates that are found in the extracellular milieu. Some of these enzymes, such as CD26, act as peptidases, while others, including CD73, CD38, CD39, ART2, and PC-1, utilize nucleotides as substrates. Although it was proposed that these nucleotide-utilizing enzymes might be involved in salvaging purines[2] or in generating products such as ATP, ADP, and adenosine that function as signaling molecules for purinergic receptors[3], until recently very little was known about the functional roles these enzymes might play during immune responses. However, in the last 10 years it has become clear that many of these enzymes play very important roles in regulating the survival, activation, and effector function of leukocytes[4]. Our laboratory has spent the last several years assessing the role of one of these ectoenzymes, CD38, in immune responses. In this article, we will review our recent work, focusing on the role that CD38 plays in regulating innate and adaptive immune responses.

From the Trudeau Institute, Saranac Lake, NY 12983, USA. Current address for S. Partidá-Sánchez, Columbus Children's Research Institute, Columbus, OH 43205, USA.

2. CD38 REGULATES INNATE AND ADAPTIVE
IMMUNE RESPONSES

CD38 is a member of a family of enzymes that catalyze NAD glycohydrolase and ADP-ribosyl cyclase reactions[5]. The different CD38 family members are structurally similar and members of the family have been identified from organisms as diverse as *Aplysia californica* (invertebrate sea slug),[6,7] *Schistosoma mansoni* (mammalian parasite),[8] and humans[9]. CD38 catalyzes the formation of three products — cyclic adenosine diphosphate ribose (cADPR), adenosine diphosphate ribose (ADPR), and nicotinic acid adenine dinucleotide (NAADP) — from its substrate(s) nicotinamide adenine dinucleotide (NAD(P))[10]. Interestingly, all three of the products generated by CD38 can induce calcium mobilization[10,11], and cADPR has been shown to regulate calcium signaling in smooth muscle, neurons, and exocrine cells[12]. However, despite the fact that CD38 is expressed on most hematopoietic cells[13, 14] and can be upregulated in response to inflammatory stimuli[15,16], it was unclear whether CD38, through its production of calcium-mobilizing metabolites, could regulate immune responses.

To address this important question, CD38-deficient mice were generated in the laboratory of Maureen Howard. The initial examination of these mice indicated that CD38 was not obligate for the development of any of the hematopoietic lineages but was necessary for optimal T cell-dependent humoral immune responses[17]. Interestingly, the CD38-deficient (CD38KO) mice were unable to produce antibody in response to vaccination when low doses of a relatively weak adjuvant (alum) were used but responded normally when immunized with protein antigens emulsified in Freund's adjuvant[17]. At the time we proposed that CD38 might function to regulate B cell activation by acting as a co-receptor for the B cell receptor (BCR). However, subsequent experiments indicated that CD38KO B cells proliferated normally in response to BCR ligation[18] and that the defective humoral immune response seen in the CD38KO mice was not due to the loss of CD38 on B lymphocytes[19]. Thus, while it was clear that CD38 did regulate humoral immune responses, we had few clues as to how CD38 or its enzymatic products might function in the immune system.

2.1. CD38 Regulates Neutrophil Migration and Lung
Inflammatory Responses

Since CD38KO mice made defective humoral immune responses when antigen dose and adjuvant were limiting[17], we considered the possibility that CD38KO mice might have difficulty in responding to the innate signals that trigger humoral immune responses. To test this hypothesis, we determined whether CD38KO mice were able to generate a normal inflammatory response. In one set of experiments, we infected CD38KO mice with the gram[+] organism *S. pneumoniae* and measured the inflammatory response. Interestingly, despite the fact that the infected CD38KO mice upregulated expression of inflammatory cytokines systemically as well as in the lung[20], the inflammatory cell infiltrate was significantly reduced in the lungs of these mice and the bacteria rapidly disseminated from the lungs to the blood[21]. The dissemination occurred within 12 hr and the majority of CD38KO mice succumbed to infection within 36 hr[20,21]. Thus, it was quite clear that these animals had profound defects in their ability to generate an innate inflammatory response after infection with S. pneumoniae.

Upon further examination, we found that neutrophils did not accumulate in the lung airways of S. pneumoniae-infected CD38KO mice[21]. This was due, at least in part, to an intrinsic defect in the CD38KO neutrophils, as we found that the chemotactic response of CD38KO neutrophils to the bacterial-derived chemoattractant, fMLF, was significantly reduced[21]. Likewise, CD38KO neutrophils were unable to migrate in response to several endogenous chemoattractants and chemokines, including serum amyloid A and MIP-1α[22]. Interestingly, CD38KO neutrophils were able to respond to the CXCR1/2 ligands, IL-8 and MIP-2[21], indicating that the cells were not refractive to all chemokines.

2.2. CD38 Regulates Dendritic Cell Trafficking in Vitro and in Vivo

Although the increased susceptibility of the CD38KO mice clearly demonstrated that, at least under some conditions, the innate inflammatory response to Streptococcus infection was dependent on CD38, we thought that it was unlikely that a migratory defect in neutrophils could explain the poor humoral immune responses observed in the CD38KO mice. Therefore, we next asked whether other cell types that respond to inflammatory stimuli were also unable to migrate normally in response to chemokines. We immediately turned our attention to dendritic cells (DCs) as these cells migrate from the blood to the peripheral tissues and from the peripheral sites to the draining lymph nodes in response to inflammation[23]. Indeed, T cell-dependent immune responses in the lymph nodes are significantly impaired in animals that lack the chemokines or chemokine receptors that induce DC migration[24,25]. To determine a potential role for CD38 in DC migration, we sort-purified immature and mature DCs from cultures of bone marrow-derived CD38KO and normal wild-type (WT) cells and tested whether these cells could migrate in chemotaxis assays. Similar to the results that we had previously obtained with CD38KO neutrophils, we found that the CD38KO DCs were intrinsically defective in their chemotactic response to an array of chemokines[19]. In particular, the immature DCs from the CD38KO bone marrow cultures were unable to migrate in response to the CXCR4 ligand CXCL12 (SDF-1) and the CCR2 ligand CCL2 (MCP-1). Likewise, the CD38KO DCs matured in vitro with TNFα did not migrate effectively in response to the CCR7 ligands CCL19 and CCL21[19] and DCs isolated from the spleen could not migrate in response to CCL19, CCL21, and CXCL12 (Figure 1). This was not due to an inability of the CD38KO DCs to mature in response to TNFα as these cells upregulated expression of costimulatory molecules like CD80 and CD86 and upregulated CCR7[19]. Instead, it again appeared that the CD38KO DCs were refractory to chemokine receptor signaling.

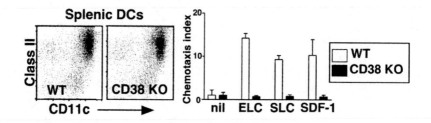

Figure 1. CD38 regulates DC chemotaxis. Dendritic cells were purified from the spleens of WT or CD38KO mice by positive selection using CD11c magnetic beads. FACS plots (left panel) indicate that the splenic DCs from both WT and CD38KO mice expressed high levels of CD11c and MHC class II. The migration of DCs to CCL19 (ELC, 10 ng/ml), CCL21 (SLC, 10 ng/ml), and CXCL12 (SDF-1, 10 ng/ml) was measured using transwell chambers. The purified DCs were placed into the upper well of the chemotaxis chamber and the number of cells that migrated to the bottom chamber within 90 minutes in response to the chemotactic stimulus was quantitated by flow cytometry. The data are reported as the mean ± SD of the chemotactic index (CI), which represents the number of cells that migrated in response to the chemokine divided by the basal migration of cells in response to control medium (nil).

We confirmed these in-vitro results using in-vivo migration assays and demonstrated that the trafficking of DC precursors from the blood to the skin was significantly impaired in CD38KO mice[19]. These results were consistent with the fact that CD38 is required for CCR2-dependent migration[19] and that CCR2 is required for trafficking of DC precursors to the skin[26]. Furthermore, we showed that DC trafficking from the skin to the lymph node after epicutaneous application of FITC was highly impaired in the CD38KO mice[19]. Again, these data were consistent with the finding that CD38 regulates the chemotaxis of DCs to CCR7 ligands and that migration of Langerhans cells in the skin to lymph nodes after FITC application is highly dependent on CCR7[24,25].

2.3. CD38 Regulates T Cell-Dependent Immune Responses

Since CD38 clearly regulated the migration of DCs both in vitro and in vivo, we predicted that the priming and activation of CD4 T cells was likely to be impaired in the CD38KO mice. To test this hypothesis, we adoptively transferred normal T cell receptor (TCR) transgenic ovalbumin (OVA)-specific CFSE-labeled T cells into CD38KO or normal WT hosts and assessed the activation, proliferation, and expansion of these cells after immunization with OVA peptide in alum. We found that the normal TCR transgenic T cells expanded upon immunization and underwent multiple rounds of proliferation in both the spleen[19] and lymph node (Figure 2). In contrast, the T cells transferred into the CD38KO hosts did not proliferate as extensively as assessed by dilution of the CFSE dye and a larger proportion of the transferred OVA-specific T cells remained CD44^{low} in the CD38KO hosts (Figure 2). These data indicated that CD38KO DCs were less effective at priming CD4 T cells, even when the T cells were able to express CD38. Since we immunized the mice with the OVA peptide, it was unlikely that the functional defect in the CD38KO DCs could simply be attributed to an inability to process and present antigen. Instead, these data argued that the intrinsic inability of the CD38KO DCs to respond to chemokines accounted, at least in part, for the poor T cell priming seen in the CD38KO mice.

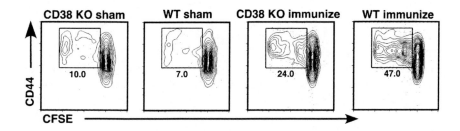

Figure 2. CD4 T cells are less efficiently primed in a CD38KO host. OVA-specific CD4 T cells were purified from the spleens and lymph nodes of Thy1.1$^+$ OT-II TCR transgenic mice, labeled with CFSE, and then transferred to either WT or CD38KO (Thy1.2$^+$) hosts (5×10^6 cells/mouse). The recipient mice were immunized i.p. with 1 μg OVA peptide in alum or with a vehicle control (sham). On day 3, the draining lymph node cells were isolated and stained with antibodies to Thy1.1, CD4, and CD44. The CFSE and CD44 profile was analyzed on the donor Thy1.1$^+$CD4$^+$ T cells. The percentage of activated CD44^{hi} donor T cells that underwent one or more rounds of cell division is indicated.

As expected, the impairment in DC trafficking and CD4 T cell priming observed in the CD38KO mice had functional consequences for the humoral immune response. First, we observed that the number of B and T cells within the lymph node was significantly reduced in the CD38KO mice that were exposed to FITC. This indicates that, in the absence of significant DC trafficking from the skin to the lymph node, lymphocytes did not accumulate or were not retained in the lymph node[19]. The number of antigen-specific CD4 T cells and B cells present in the lymph node at 1 week after vaccination was also significantly reduced in the CD38KO mice, and antigen-specific antibody was reduced by at least 10-fold, regardless of whether the mice were immunized systemically with protein antigen or at a local site with FITC[19]. Taken altogether, our data showed that CD38 regulates the trafficking of multiple cell types, including neutrophils, monocytes, and dendritic cells in response to vaccination and infection and that this has important consequences for the initiation of both innate and adaptive immune responses.

3. CD38 MODULATES CHEMOKINE RECEPTOR SIGNALING BY PRODUCING CALCIUM MOBILIZING METABOLITES

While our data clearly demonstrated that CD38 regulates the chemotactic response of leukocytes to an array of chemokines and chemoattractants, we still did not understand how CD38 was controlling cell migration or whether the enzyme activity of CD38 was necessary for its function. However, in elegant studies from a number of other labs, it had become clear that cADPR, one of the metabolites generated in the CD38 enzyme reaction, was able to mobilize calcium release from ryanodine receptor-gated stores in smooth muscle cells, neurons, and exocrine cells that were stimulated with ligands of a number of different G-protein coupled receptors (GPCRs)[27]. Since chemokine receptors are also members of the GPCR family, and leukocytes are reported to express some ryanodine receptor isoforms[28], we considered the possibility that CD38 might mediate calcium mobilization by producing cADPR in chemokine-stimulated cells. As expected, hematopoietic cells expressing CD38 were competent to produce cADPR when NAD, the CD38 substrate, was present in the media[21]. Furthermore, we found that hematopoietic cells could mobilize calcium when exposed to cADPR and that this calcium release could be blocked with a specific cADPR antagonist, 8Br-cADPR, as well as with ryanodine receptor antagonists[21]. These data indicated that hematopoietic cells express all of the machinery necessary to mediate cADPR-dependent calcium release.

Table 1. Chemokine Receptors Regulated by CD38 or cADPR

Chemokine receptors	Mouse/human	Cell type	CD38 dependent	cADPR dependent
CCR1	Mouse	Neutrophil	Yes	Yes
CCR1	Human	Monocyte	?	Yes
CCR2	Mouse	Immature DC	Yes	Yes
CCR5	Human	Monocyte	?	Yes
CCR7	Mouse	Mature DC	Yes	Yes
CXCR1/2	Mouse and human	Neutrophil	No	No
CXCR4	Mouse	Immature DC	Yes	Yes
CXCR4	Human	Monocyte	?	Yes
FPRL1	Mouse and human	Neutrophil	Yes	Yes
FPR1	Mouse	Neutrophil	Yes	Yes
High-affinity FPR	Human	Neutrophil	No	No

When we examined the calcium response of chemokine-stimulated CD38KO neutrophils or DCs, we found that the response was greatly attenuated. Specifically, we found that the chemokine-treated CD38KO cells were not able to make a sustained calcium response and that the influx of extracellular calcium into the chemokine-stimulated CD38KO cells was almost entirely ablated[21,22]. Likewise, the calcium response to an array of chemokines was defective in WT neutrophils and DCs that were first pretreated with the specific cADPR antagonist. In fact, these responses were indistinguishable from the calcium response of CD38KO cells[19,21]. Perhaps most importantly, WT cells that were pretreated with the cADPR antagonist were unable to migrate in response to a number of chemoattractants including the ligands for FPRL1, CCR1, CCR2, CXCR4, and CCR7 (summarized in Table 1). Interestingly, and in agreement with our earlier observations using CD38KO neutrophils, 8Br-cADPR-treated WT cells were still competent to respond to CXCR1/2 ligands[21]. Together, these data argued very strongly that cADPR produced by CD38 does regulate calcium signaling in chemokine-stimulated leukocytes and, more importantly, also regulates the chemotactic response of these cells to a specific and discrete subset of chemokines and chemoattractants.

We have since extended this work to test whether cADPR antagonists can be used to block the calcium and chemotactic response of human neutrophils and monocytes to inflammatory chemokines and found that cADPR antagonists can act as very potent inhibitors of human leukocyte migration at least in vitro[22] (Figure 3). Similar to what we observed with the mouse leukocytes, we found that the 8Br-cADPR-treated human leukocytes were still competent to migrate to some chemoattractants including IL-8 and the HIV T-20 peptide (Figure 3), indicating that cADPR regulates signaling through a select subset of chemokine receptors in both mouse and human cells. Since the cADPR antagonist was effective in blocking migration of leukocytes to some inflammatory chemokines, we postulated that CD38 enzyme inhibitors might make effective anti-inflammatory drugs. Unfortunately, no small molecule inhibitors of the CD38 enzyme reaction have been identified to date. However, it has been reported that some NAD analogues act as potent competitive antagonists of CD38 glycohydrolase and cyclase activity[29], suggesting that NAD analogues might be effective in vivo. To begin to test this possibility, we treated mouse and human leukocytes with the NAD analogue N(8-Br-A)D$^+$, reasoning that if the cells expressed CD38 they would produce the cADPR antagonist, 8Br-cADPR, from N(8-Br-A)D$^+$. Excitingly, treatment of leukocytes with the NAD analogue very effectively attenuated the calcium response of these cells to FPRL1, CCR1, CCR2, CXCR4, and CCR7 ligands and completely blocked their migration to these chemokines in vitro[19,21,22] (Figure 3). Thus, we are confident that CD38 substrate analogues or cADPR antagonists may be useful in blocking inflammatory responses and hope to test this hypothesis soon.

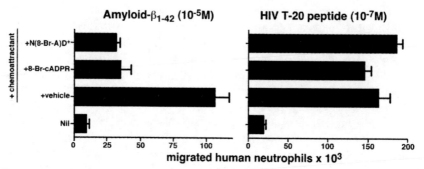

Figure 3. A CD38 substrate analogue and a cADPR antagonist block the chemotaxis of human peripheral blood neutrophils to FPRL-1 ligands but not FPR ligands. Human peripheral blood neutrophils isolated from normal donors were preincubated in medium (nil) or in medium containing 8-Br-cADPR (100 μM) or N(8-Br-A)D$^+$ (500 μM) for 15 minutes and then placed in the upper wells of chemotaxis chambers. The lower wells of the chambers contained either the FPRL-1 ligand Amyloid-β_{1-42} peptide (10^{-5} M) or the high-affinity FPR ligand HIV T-20 peptide (10^{-7} M). The cells that migrated to the bottom chamber in 45 minutes were collected and enumerated by flow cytometry. The mean ± SD of triplicate wells is shown.

4. CONCLUSIONS AND FUTURE DIRECTIONS

In summary, we have shown that at least one of the ectoenzymes expressed by mouse and human leukocytes does play an important role in regulating immune responses. However, instead of salvaging extracellular purines or inducing purinergic receptor signaling, it appears that one major function of CD38 is to produce the calcium-mobilizing metabolite cADPR from its substrate, extracellular NAD. This metabolite regulates calcium responses in monocytes, neutrophils, and dendritic cells activated with a discrete subset of chemoattractants and controls the migration of these cells to sites of inflammation and secondary lymphoid tissues. In the absence of CD38-dependent cell trafficking, both innate and adaptive immune responses are significantly attenuated. Perhaps most important from a clinical perspective, CD38 substrate analogues and cADPR antagonists can be used to block the migration of human leukocytes to many, but not all, inflammatory chemokines.

4.1. Unresolved Questions

Despite the significant progress we have made in understanding the role of the ecto-enzyme CD38 in immune responses, there are a number of remaining unanswered questions. For example, we do not yet know why CD38 and cADPR regulate signaling through some chemokine receptors and not others. However, our preliminary experiments suggest that chemokine receptors can be subdivided depending on which class of G proteins associate with the receptor and whether the calcium response is CD38 and cADPR dependent. We also do not yet know whether NAADP and ADPR, the other calcium-mobilizing metabolites produced by CD38, are important for chemokine receptor signaling, nor do we have a good understanding of how any of the CD38-generated metabolites are transported from outside the DC or neutrophil into the cytosol. Finally, we do not yet know whether CD38 plays any non-enzyme-dependent roles in immune responses. This question is particularly important, as we and others have shown that CD38 can also function as a receptor on lymphocytes[18,30]. Human CD38 has been shown to bind CD31[31,32], which is prominently expressed by endothelial cells, and both mouse and human CD38 can bind to hyaluronic acid[33,34], a component of extracellular matrix. Thus, one can envision that CD38 might not only regulate the chemotaxis of cells via an enzyme-dependent mechanism but could also mediate adhesion of the cells within tissues and blood vessels. Since the structure of CD38 has been elucidated[35], and the key catalytic residues have been identified[36], we now hope to identify other non-enzyme-dependent roles for CD38 by making animals that express enzymatically inactive CD38.

4.2. Model

Perhaps the most intriguing question remaining to be addressed is why NAD, a dinucleotide that is usually found almost exclusively in cells and is critically important for ATP generation in the mitochondria, would be present outside the cell and why catabolism of this nucleotide would be important for immune cell function. We hypothesize that extracellular NAD might function to activate innate immune mechanisms (Figure 4). If this is correct, then we expect that extracellular NAD levels will rise in inflamed or damaged tissues and that CD38-expressing cells will catabolize this free NAD. Indeed, it is known that ecto-NAD levels are very low in the serum of normal mice[37], suggesting that under homeostatic conditions very little free NAD is available. Interestingly, in CD38KO mice, serum, and tissue NAD levels are elevated[38], indicating that NAD is released or actively transported to extracellular sites and that CD38 normally catabolizes the "free" ecto-NAD. Upon inflammation or damage, local levels of extracellular NAD can rise quite dramatically, particularly in mice that lack CD38, and cannot efficiently catabolize the free NAD[38]. Therefore, we propose that NAD is released by dying cells during infection or damage and that, once NAD is available, CD38-expressing cells will catabolize the free NAD and produce cADPR. Cyclic ADP-ribose producing cells respond to chemokine receptor ligation by making a prolonged calcium response and are then competent to migrate to inflamed tissues and lymphoid tissues. Thus, one of the major functions of CD38 might be to "sense" elevated extracellular NAD and to produce metabolites from the ecto-NAD that can enhance the migratory capacity of leukocytes at times of stress or damage. This would put CD38 in the same category as other damage and pathogen "sensing" receptors such as the Toll-like receptors (TLRs) that are activated by pathogen components and endogenous heat shock proteins[30]. It further suggests that CD38, like the TLRs, stands firmly at the intersection of innate and adaptive immune responses and that therapies that are designed to either augment or suppress the CD38 and ecto-NAD interaction may prove useful as immune modulators.

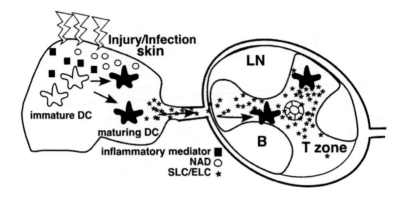

Figure 4. CD38 regulates innate and adaptive immune responses by "sensing" extracellular NAD released in infected or damaged tissues. We propose in our current working model that CD38 binds extracellular NAD released by damaged or dying cells and produces calcium-mobilizing metabolites like cADPR. The cADPR produced by the CD38-expressing cells enhances the ability of these cells to migrate in response to certain chemokines. We further propose that compounds that alter NAD levels or CD38 enzyme activity could be used to block inflammatory responses.

5. ACKNOWLEDGMENTS

The authors would like to thank Dr. Troy Randall for his comments and helpful discussions during the course of this work. We gratefully acknowledge financial support for this work from NIH grants AI-43629 and AI-057996, and the Trudeau Institute.

6. REFERENCES

1.   J.L. Cannons and P.L. Schwartzberg. Fine-tuning lymphocyte regulation: what's new with tyrosine kinases and phosphatases? *Curr Opin Immunol* **16**:296–303 (2004).
2.   P. Deterre, L. Gelman, H. Gary-Gouy, C. Arrieumerlou, V. Berthelier, J.-M. Tixier, S. Ktorza, J. Goding, C. Schmitt and G. Bismuth. Coordinated regulation in human T cells of nucleotide-hydrolyzing ecto-enzymatic activities, including CD38 and PC-1. *J Immunol* **157**:1381–1388 (1996).
3.   R. Resta, Y. Yamashita and L.F. Thompson. Ecto-enzyme and signaling functions of lymphocyte CD73. *Immunol Rev* **161**:95–109 (1998).

4. M. Salmi and S. Jalkanen. Cell-surface enzymes in control of leukocyte trafficking. *Nat Rev Immunol* **5**:760–771 (2005).

5. F. Schuber and F.E. Lund. Structure and enzymology of ADP-ribosyl cyclases: conserved enzymes that produce multiple calcium mobilizing metabolites. *Curr Mol Med* **4**:249–261 (2004).

6. H.C. Lee and R. Aarhus. ADP-ribosyl cyclase: an enzyme that cyclizes NAD⁺ into a calcium mobilizing metabolite. *Cell Regul* **2**:203–209 (1991).

7. M.R. Hellmich and F. Strumwasser. Purification and characterization of a molluscan egg-specific NADase, a second-messenger enzyme. *Cell Regul* **2**:193–202 (1991).

8. S.P. Goodrich, H. Muller-Steffner, A. Osman, M.J. Moutin, K. Kusser, A. Roberts, D.L. Woodland, T.D. Randall, E. Kellenberger, P.T. LoVerde, F. Schuber and F.E. Lund. Production of calcium-mobilizing metabolites by a novel member of the ADP-ribosyl cyclase family expressed in Schistosoma mansoni. *Biochemistry* **44**:11082–11097 (2005).

9. D.G. Jackson and J.I. Bell. Isolation of a cDNA encoding the human CD38 (T10) molecule, a cell surface glycoprotein with an unusual discontinuous pattern of expression during lymphocyte differentiation. *J Immunol* **144**:2811–2815 (1990).

10. H.C. Lee. Multiplicity of Ca2+ messengers and Ca2+ stores: a perspective from cyclic ADP-ribose and NAADP. *Curr Mol Med* **4**:227–237 (2004).

11. F.J. Kuhn, I. Heiner and A. Luckhoff. TRPM2: a calcium influx pathway regulated by oxidative stress and the novel second messenger ADP-ribose. *Pflugers Arch* **451**:212–219 (2005).

12. A.H. Guse. Regulation of calcium signaling by the second messenger cyclic adenosine diphosphoribose (cADPR). *Curr Mol Med* **4**:239–248 (2004).

13. J.E. Fernandez, S. Deaglio, D. Donati, I.S. Beusan, F. Corno, A. Aranega, M. Forni, B. Falini and F. Malavasi. Analysis of the distribution of human CD38 and of its ligand CD31 in normal tissues. *J Biol Reg Homeost Agents* **12**:81–91 (1998).

14. F. Malavasi, A. Funaro, M. Alessio, L.B. De Monte, C.M. Ausiello, U. Dianzani, F. Lanza, E. Magrini, M. Momo and S. Roggero. CD38: a multi-lineage cell activation molecule with a split personality. *Int J Clin Lab Res* **22**:73–80 (1992).

15. T. Musso, S. Deaglio, L. Franco, L. Calosso, R. Badolato, G. Garbarino, U. Dianzani and F. Malavasi. CD38 expression and functional activities are up-regulated by IFN-g on human monocytes and monocytic cell lines. *J Leukoc Biol* **69**:605–612 (2001).

16. B. Bauvois, L. Durant, J. Laboureau, E. Barthelemy, D. Rouillard, G. Boulla and P. Deterre. Upregulation of CD38 gene expression in leukemic B cells by interferon types I and II. *J Interferon Cytokine Res* **19**:1059–1066 (1999).

17. D. Cockayne, T. Muchamuel, J.C. Grimaldi, H. Muller-Steffner, T.D. Randall, F.E. Lund, R. Murray, F. Schuber and M.C. Howard. Mice deficient for the ecto-NAD⁺ glycohydrolase CD38 exhibit altered humoral immune responses. *Blood* **92**:1324–1333 (1998).

18. F.E. Lund, D.A. Cockayne, T.D. Randall, N. Solvason, F. Schuber and M.C. Howard. CD38: A new paradigm in lymphocyte activation and signal transduction. *Immunol Rev* **161**:79–93 (1998).

19. S. Partida-Sanchez, S. Goodrich, K. Kusser, N. Oppenheimer, T.D. Randall and F.E. Lund. Regulation of dendritic cell trafficking by the ADP-ribosyl cyclase CD38: Impact on the development of humoral immunity. *Immunity* **20**:279–291 (2004).

20. S. Partida-Sanchez, T.D. Randall and F.E. Lund. Innate immunity is regulated by CD38. an ecto-enzyme with ADP-ribosyl cyclase activity. *Microbes Infect* **5**:49–58 (2003).

21. S. Partida-Sanchez, D.A. Cockayne, S. Monard, E.L. Jacobson, N. Oppenheimer, B. Garvy, K. Kusser, S. Goodrich, M. Howard, A. Harmsen, T.D. Randall and F.E. Lund. Cyclic ADP-ribose production by CD38 regulates intracellular calcium release, extracellular calcium influx and chemotaxis in neutrophils and is required for bacterial clearance in vivo. *Nat Med* **7**:1209–1216 (2001).

22. S. Partida-Sanchez, P. Iribarren, M.E. Moreno-Garcia, J.-L. Gao, P.M. Murphy, N. Oppenheimer, J.M. Wang and F.E. Lund. Chemotaxis and calcium responses of phagocytes to formyl-peptide receptor ligands is differentially regulated by cyclic ADP-ribose. *J Immunol* **172**:1896–1906 (2004).

23. G.J. Randolph, G. Sanchez-Schmitz and V. Angeli. Factors and signals that govern the migration of dendritic cells via lymphatics: recent advances. *Springer Semin Immunopathol* **26**:273–287 (2005).

24. M.D. Gunn, S. Kyuwa, C. Tam, T. Kakiuchi, A. Matsuzawa, L.T. Williams and H. Nakano. Mice lacking expression of secondary lymphoid organ chemokine have defects in lymphocyte homing and dendritic cell localization. *J Exp Med* **189**:451–460 (1999).

25. R. Forster, A. Schubel, D. Breitfeld, E. Kremmer, I. Renner-Muller, E. Wolf and M. Lipp. CCR7 coordinates the primary immune response by establishing functional microenvironments in secondary lymphoid organs. *Cell* **99**:23–33 (1999).

26. M. Merad, M.G. Manz, H. Karsunky, A. Wagers, W. Peters, I. Charo, I.L. Weissman, J.G. Cyster and E.G. Engleman. Langerhans cells renew in the skin throughout life under steady-state conditions. *Nat Immunol* **3**:1135–1141 (2002).

27. H.C. Lee. Physiological functions of cyclic ADP-ribose and NAADP as calcium messengers. *Annu Rev Pharmacol Toxicol* **41**:317–345 (2001).

28. E. Hosoi, C. Nishizaki, K.L. Gallagher, H.W. Wyre, Y. Matsuo and Y. Sei. Expression of the ryanodine receptor isoforms in immune cells. *J Immunol* **167**:4887–4894 (2001).

29. H. Muller-Steffner, O. Malver, N.J. Oppenheimer and F. Schuber. Slow-binding inhibition of NAD^+ glycohydrolase by arabino analogues of b-NAD^+. *J Biol Chem* **267**:9606–9611 (1992).

30. S. Deaglio and F. Malavasi. Human CD38: a receptor, an (ecto) enzyme, a disease marker and lots more. *Mod Asp Immunobiol* **2**:121–125 (2002).

31. S. Deaglio, U. Dianzani, A.L. Horenstein, J.E. Fernandez, C. van Kooten, M. Bragardo, A. Funaro, G. Garbarino, F. Di Virgilio, J. Bancherau and F. Malavasi. Human CD38 ligand: a 120-kDA protein predominantly expressed on endothelial cells. *J Immunol* **156**:727–734 (1996).

32. S. Deaglio, M. Morra, R. Mallone, C.M. Ausiello, E. Prager, G. Garbarino, U. Dianzani, H. Stockinger and F. Malavasi. Human CD38 (ADP-Ribosyl Cyclase) is a counter-receptor of CD31, an Ig superfamily member. *J Immunol* **160**:395–402 (1998).

33. F.E. Lund, T.D. Randall and S. Partida-Sanchez. Cyclic ADP-ribose and NAADP. In *Structures, metabolism and function*, Vol. 1, pp. 217–240. Ed. H.C. Lee. New York: Kluwer Academic (2002).

34. H. Nishina, K. Inageda, K. Takahashi, S. Hoshino, K. Ikeda and T. Katada. Cell surface antigen CD38 identified as ecto-enzyme of NAD glycohydrolase has hyaluronate-binding activity. *Biochem Biophys Res Comm* **203**:1318–1323 (1994).

35. Q. Liu, I.A. Kriksunov, R. Graeff, C. Munshi, H.C. Lee and Q. Hao. Crystal structure of human CD38 extracellular domain. *Structure (Cambridge)* **13**:1331–1339 (2005).

36. C. Munshi, R. Aarhus, R. Graeff, T.F. Walseth, D. Levitt and H.C. Lee. Identification of the enzymatic active site of CD38 by site-directed mutagenesis. *J Biol Chem* **275**:21566–21571 (2000).

37. U.H. Kim, M.K. Kim, J.S. Kim, M.K. Han, B.H. Park and H.R. Kim. Purification and characterization of NAD glycohydrolase from rabbit erythrocytes. *Arch Biochem Biophys* **305**:147–152 (1993).

38. M. Seman, S. Adriouch, F. Haag and F. Koch-Nolte. Ecto-ADP-ribosyltransferases (ARTs): emerging actors in cell communication and signaling. *Curr Med Chem* **11**:857–872 (2004).

39. G.M. Barton and R. Medzhitov. Toll-like receptors and their ligands. *Curr Top Microbiol Immunol* **270**:81–92 (2002).

13

VASCULAR LEUKOCYTES: A POPULATION WITH ANGIOGENIC AND IMMUNOSSUPPRESSIVE PROPERTIES HIGHLY REPRESENTED IN OVARIAN CANCER

George Coukos, Jose R. Conejo-Garcia,
Ron Buckanovich, and Fabian Benencia

1. PHYSIOLOGICAL ANGIOGENESIS VS. PATHOLOGICAL ANGIOGENESIS

Physiological angiogenesis occurs mainly during the embryonic stage, originating a vascular network that meets the nutritional and functional demands of the developing organism. In the embryo, formation of blood vessels occurs via vasculogenesis and angiogenesis. Vasculogenesis involves the de-novo differentiation of endothelial cells from angioblasts, mesoderm-derived precursor cells, which assemble into primary capillary vessels[1]. This network differentiates then by angiogenesis, where new vessels arise from sprouting of preexisting capillaries. In the adult, physiological neovessel formation is involved in wound healing, tissue remodeling, and the female reproductive cycle, while pathological angiogenesis is associated with ischemia, rheumatoid arthritis, diabetic retinopathy, age-related macular degeneration, psoriasis, inflammatory bowel diseases, endometriosis, and tumor neovascularization[2,3]. Pathological neovascularization is characterized by increased vascular permeability, which leads to leakage, hemorrhaging, and inflammation. Although sprouting of blood vessels is the principal process in neovascularization, other mechanisms such as intussusception or cooptation of circulating endothelial cell progenitors have also been described[4].

Center for Research in Reproduction and Women's Health; and Abramson Family Cancer Research Institute, University of Pennsylvania, BRBII/III, 421 Curie Blvd, Philadelphia, PA19104.

2. ENDOTHELIAL PROGENITORS AND NEOANGIOGENESIS

It has been proposed that cooption of endothelial cell progenitors is an additional mechanism of neoangiogenesis. The existence of this angioblast-like circulating cell has been shown in certain murine models[5]. Moreover, it seems that the recruitment and in-situ differentiation of bone marrow-derived endothelial progenitor cells is essential to promote effective neovascularization[6,7]. Particularly, several reports suggested that adult bone marrow contains a scarce population of endothelial progenitor cells that can be mobilized to the circulation and contribute to neovascularization. In murine models of bone marrow transplantation, donor bone marrow cells were found to incorporate into vessels during wound healing, ischemia, corneal neovascularization, and tumor growth upon transplantation into recipient wild-type or immunodeficient animals[8,9]. These data indicated that circulating endothelial cell progenitors are mobilized and contribute to different events in postnatal neovascularization and regenerative processes. Although the exact mechanisms by which endothelial cell precursors are unknown, it has been showed that high levels of VEGF are responsible for recruitment of endothelial cell precursors to the tumor, where they incorporate into neovessels[10].

3. HEMATOPOIETIC CELLS PARTICIPATE IN NEOANGIOGENESIS

The participation of cells other than typical endothelial cells during the process of neoangiogenesis has become the focus of numerous recent studies. Several populations of hematopoietic cells assume an endothelial phenotype when cultured under pro-angiogenic conditions. These include $CD34^+$, $Sca1^+$, $CD133^+$, and $CD14^+$ cells. Endothelial cell progenitors were first identified by expression of the hematopoietic stem cell antigens, CD34 and flk-1, by other hematopoietic stem cell antigens, such as CD133 (AC133)[5]. More recently different studies have demonstrated that monocytes or monocyte-like cells can also function as endothelial cell progenitors and incorporate into growing vasculature in experimental models[11].

4. ANTIGEN-PRESENTING CELLS AS ENDOTHELIAL CELLS

Antigen-presenting cells (APCs) such as monocytes/macrophages and dendritic cells have the capability to phagocyte antigen, process it, and present it to T cells in the context of major histocompatibility complex in order to initiate an immune response. APCs mature after receiving the antigenic stimuli, increasing the levels of expression of MHC and costimulatory molecules, thus ensuring proper stimuli for T cell activation. In recent years an additional capability of APCs has been reported. It has been shown that monocytes or dendritic cells cultured in the presence of angiogenic factors such as VEGF undergo an endothelization process characterized by the lost of CD14/CD45 and displayed endothelial markers CD31, CD34 von Willebrand factor (vWF), vascular endothelial growth factor receptor-2 (VEGFR-2), and VE-cadherin. They displayed other characteristics of endothelial cells, such as LDL uptake, lectin binding, and formation of cord-like structures in 3D gels. Several reports have shown that these cells are able to assemble into vascular structures in vitro and in vivo[12-16]. Furthermore, it has recently been reported that these cells even acquire functional properties similar to brain microvascular endothelial cells under the appropriate stimuli[17]. On the other hand, treatment of the aforementioned cells with inflammatory molecules render typical antigen-presenting cells with the capability to activate T cells[18].

5. TUMOR ANGIOGENESIS

It has been proposed that dormant avascular tumor nodules could only grow and develop if they become vascularized[19]. Induction of this "angiogenic switch" on tumors involves a change in the balance between various molecules with the capability to activate or inhibit vascular growth. These molecules include basic fibroblast growth factor, epidermal growth factor, platelet-derived growth factor, matrix metalloproteinases, placental growth factor, and angiopoietins-1 and -2, among others [20]. In particular, a key molecule in tumor angiogenesis activators is the vascular endothelial growth factor (VEGF). VEGF alone can initiate the angiogenic cascade of events, acting as a proliferation, migration, and survival factor for endothelial cells, both in vitro and in vivo[21]. Thus, VEGF induces new blood vessel formation in the tumors, which supplies tumor cells with an adequate supply of oxygen, metabolites, and an effective way to remove waste products. VEGF is secreted by the vast proportion of human cancers[21], and we have reported that high levels of VEGF in human ovarian carcinoma are related to poor prognosis[22]. Moreover, in a mouse tumor model of ovarian carcinoma developed in our laboratory we showed that VEGF overexpression highly increases tumor vascularization[23].

6. VASCULAR LEUKOCYTES

We have recently reported, in human ovarian carcinoma, the presence of a subset of CD45$^+$/CD11c$^+$ dendritic cells expressing high levels of endothelial markers such as VE–Cadherin, CD31, CD34, and CD146[24]. These cells constitute a high proportion of the CD45 population in human ovarian cancer (20–60%) in most of the samples analyzed (Figure 1A). Phenotypical characterization of this population also showed expression of MHC-II, and low levels of costimulatory molecules CD80 and CD86. By means of cell sorting we were able to obtain a highly purified population of these cells, and we showed that they have the capacity to assemble into endothelial-like structures in vitro and to create per fusible blood vessels in vivo (Figure 1B). These cells were also detected in tumoral tissue by means of immunofluorescense analysis (Figure 1C). Taking into account the dual nature of these cells, we named them vascular leukocytes (VLCs).

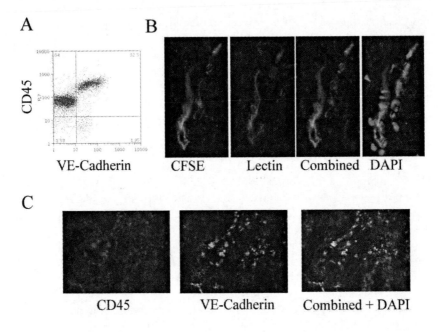

Figure 1. Angiogenic capability of human VLCs. (A) Population of leukocyte expression endothelial markers was observed in human ovarian carcinomas. These cells, named vascular leukocytes (VLCs), were recovered from ovarian carcinomas stained with CFSE and injected into the flank of NOD/SCID mice admixed in Matrigel (B). Ten days later, biotinylated tomato lectin was administered intravenously in order to stain functional vessels and Matrigel plugs were recovered for histological analysis. Samples were stained with streptavidin-rhodamine and DAPI. (C) Immunofluorescence analysis showing co-staining of CD45 and VE–Cadherin in human tumors.

In order to further study these cellular populations, we generated a mouse model of ovarian carcinoma co-expressing VEGF, a molecule reportedly involved in the endothelization of leukocytes, and β-defensin 29, an antimicrobial peptide that chemoattracts immature DCs through CCR6[25]. By recruiting DCs to tumor sites in the presence of high levels of VEGF we expected to recapitulate the microenvironmental conditions from which human VLCs are recovered. In our experimental syngeneic model we were able to increase the levels of CD11c$^+$ DCs in the CD45$^+$ peritoneal population from 2–4% up to 60% in the ascites model (Figure 2A). More than 90% of CD11c$^+$CD45$^+$ cells were MHC-II$^+$, but they expressed very low levels of costimulatory molecules, and were thus identified as DC precursors. In the flank model, these cells localized to the luminal

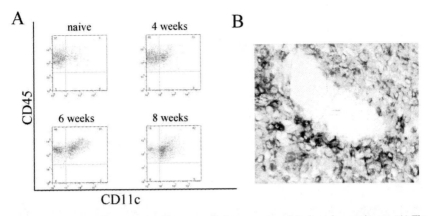

Figure 2. Dendritic cells accumulate intratumorally in a mouse model of ovarian carcinoma. (A) The proportion of CD11c cells dramatically increases in the peritoneal cavity of mice developing ascites in a mouse model of ovarian carcinoma coexpressing high levels of VEGF A and beta-defensin, an attractor for immature DCs. (B) Positive CD11c staining was observed associated to endothelial structures in solid tumors of beta-defensin-VEGF mouse ovarian carcinoma.

surface of capillary-like structures, indicating that they had contributed to the increased vascular density (Figure 2B). These cells expressed high levels of endothelial markers, such as CD31, CD34, VE-Cadherin, CD146, and VEGFR1 and 2, thus being the mouse counterpart to human VLCs (Figure 3). Moreover, VLCs transplanted in Matrigel plugs in vivo assembled into vascular-like structures, with the capability to transport blood as determined by intravascular perfusion of fluorescent dextran (Figure 4). By means of immunofluorescense analysis, we showed that these neovessels retain the expression of CD11c and CD45[25].

Furthermore, we were able to reproduce this endothelization process in vitro. We obtained highly pure murine bone marrow-derived DCs. These CD45$^+$CD11c$^+$CD34$^-$ cells showed typical characteristics of DCs and were able to induce antitumor immune responses when used as vaccines in different mouse tumor models[26]. Upon incubation with media conditioned by tumor cells expressing high levels of VEGF tumor, these cells overexpressed endothelial markers, CD31, CD34, and vWF and were able to assemble into vascular structures in vitro. Moreover, in-vitro obtained VLCs exhibited typical endothelial cell features such as Weibel-Palade bodies and endocytic vesicles. In addition, they created intercellular junctions, organizing themselves around a lumen, which are typical morphological properties of endothelial cells[26]. VEGF, important for the survival and proliferation of bona-fide endothelial cells, was the critical factor for endothelization since blocking VEGFR-2 (but not other VEGF receptors) with neutralizing antibodies stopped the transdifferentiation process.

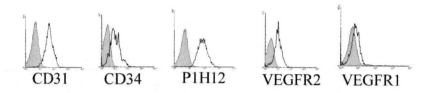

Figure 3. Vascular leukocytes are a subpopulation of DCs that express endothelial markers and low levels of costimulatory molecules. CD11c+/VE–Cadherin+/CD45+ cells recovered from beta-defensin–VEGF mouse ovarian carcinomas were analyzed for different markers of endothelial cells.

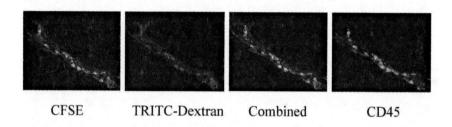

Figure 4. Angiogenic capability of mouse VLCs. VLCs recovered from beta-defensin–VEGF mouse ovarian carcinomas were stained with CFSE and injected into the flank of syngeneic mice admixed in Matrigel. Ten days later TRITC-coupled dextran was administered intravenously in order to stain functional vessels and Matrigel plugs were recovered for histological analysis.

7. VASCULAR LEUKOCYTES AND ANTITUMOR IMMUNE RESPONSE

The presence of antigen-presenting cells with immunosuppressive properties within the tumor microenvironment has been widely reported. In particular, dendritic cells showing low levels of costimulatory molecules have been reported within tumor microenvironments characterized by high levels of VEGF[27,28]. These tumor-associated APCs show highly immunosuppressive properties, rendering anergic or tolerized T cells, thus abrogating antitumor immune responses. Tumor-associated DCs have been shown to overcome this tolerogenic status after treatment with inflammatory molecules. Similarly, we have reported that VLCs express very low levels of costimulatory molecules, while producing high levels of VEGF[25]. This status can be reversed by incubation with inflammatory molecules such as LPS, TNF, and CPGs, rendering the cells capable of inducing specific T cell responses [25].

8. FINAL REMARKS

VLCs comprise a population of $CD45^+$/Vecadherin$^+$ with the dual capability of presenting antigen or assembling into endothelial cell structures. These cells are present in higher proportions of the $CD45^+$ population in human ovarian carcinomas, and in a syngeneic mouse model of ovarian cancer overexpressing VEGF. Within the tumor microenvironment, this dual function contributes to tumor growth by increasing neoangiogenesis or silencing antitumor immune responses. Thus, this population is an interesting target for antitumor therapies, since depletion of VLCs may impair neoangiogenesis within the tumor.

9. ACKNOWLEDGMENTS

This work was supported by NIH R01 CA098951 and NCI ovarian SPORE P01-CA83638. F.B. was supported by NIH D43 TW00671 funded by the Fogarty International Center.

10. REFERENCES

1. S. Patan. Vasculogenesis and angiogenesis as mechanisms of vascular network formation, growth and remodeling. *J Neurooncol* **50**(1–2):1–15 (2000).
2. M. Papetti and I.M. Herman. Mechanisms of normal and tumor-derived angiogenesis. *Am J Physiol Cell Physiol* **282**(5):C947–C970 (2002).
3. N. Ferrara. VEGF as a therapeutic target in cancer. *Oncology* **69**(Suppl 3):11–16 (2005).
4. V. Djonov, O. Baum and P.H. Burri. Vascular remodeling by intussusceptive angiogenesis. *Cell Tissue Res* **314**(1):107–117 (2003).
5. S. Rafii. Circulating endothelial precursors: mystery, reality, and promise. *J Clin Invest* **105**(1):17–19 (2000).
6. C. Urbich and S. Dimmeler. Endothelial progenitor cells functional characterization. *Trends Cardiovasc Med* **14**(8):318–322 (2004).
7. C. Urbich and S. Dimmeler. Endothelial progenitor cells: characterization and role in vascular biology. *Circ Res* **95**(4):343–353 (2004).
8. T. Asahara, H. Masuda, T. Takahashi, C. Kalka, C. Pastore, M. Silver, M. Kearne, M. Magner and J.M. Isner. Bone marrow origin of endothelial progenitor cells responsible for postnatal vasculogenesis in physiological and pathological neovascularization. *Circ Res* **85**(3):221–228 (1999).
9. Y. Lin, D.J. Weisdorf, A. Solovey and R.P. Hebbel. Origins of circulating endothelial cells and endothelial outgrowth from blood. *J Clin Invest* **105**(1):71–77 (2000).
10. D. Lyden, K. Hattori, S. Dias, C. Costa, P. Blaikie, L. Butros, A. Chadburn, B. Heissig, W. Marks, L. Witte, Y. Wu, D. Hicklin, Z. Zhu, N.R. Hackett, R.G. Crystal, M.A. Moore, K.A. Hajjar, K. Manova, R. Benezra and S. Rafii. Impaired recruitment of bone-marrow-derived endothelial and hematopoietic precursor cells blocks tumor angiogenesis and growth. *Nat Med* **7**(11):1194–1201 (2001).

11. J. Rehman, J. Li, C.M. Orschell and K.L. March. Peripheral blood "endothelial pro-genitor cells" are derived from monocyte/macrophages and secrete angiogenic growth factors. *Circulation* **107**(8):1164–1169 (2003).

12. B. Fernandez Pujol, F.C. Lucibello, U.M. Gehling, K. Lindemann, N. Weidner, M.L. Zuzarte, J. Adamkiewicz, H.P. Elsasser, R. Muller and K. Havemann. Endo-thelial-like cells derived from human CD14 positive monocytes. *Differentiation* **65**(5):287–300 (2000).

13. B. Fernandez Pujol, F.C. Lucibello, M. Zuzarte, P. Lutjens, R. Muller and K. Have-mann. Dendritic cells derived from peripheral monocytes express endothelial mark-ers and in the presence of angiogenic growth factors differentiate into endothelial-like cells. *Eur J Cell Biol* **80**(1):99–110 (2001).

14. A. Schmeisser, C.D. Garlichs, H. Zhang, S. Eskafi, C. Graffy, J. Ludwig, R.H. Strasser and W.G. Daniel. Monocytes coexpress endothelial and macrophagocytic lineage markers and form cord-like structures in Matrigel under angiogenic condi-tions. *Cardiovasc Res* **49**(3):671–680 (2001).

15. L. Yang, L.M. DeBusk, K. Fukuda, B. Fingleton, B. Green-Jarvis, Y. Shyr, L.M. Matrisian, D.P. Carbone and P.C. Lin. Expansion of myeloid immune suppressor Gr+CD11b+ cells in tumor-bearing host directly promotes tumor angiogenesis. *Cancer Cell* **6**(4):409–421 (2004).

16. Y. Zhao, D. Glesne and E. Huberman. A human peripheral blood monocyte-derived subset acts as pluripotent stem cells. *Proc Natl Acad Sci USA* **100**(5):2426–2431 (2003).

17. J. Glod, D. Kobiler, M. Noel, R. Koneru, S. Lehrer, D. Medina, D. Maric and H.A. Fine. Monocytes form a vascular barrier and participate in vessel repair after brain injury. *Blood* **107**(3):940–946 (2006).

18. A.P. Vicari, C. Chiodoni, C. Vaure, S. Ait-Yahia, C. Dercamp, F. Matsos, O. Reynard, C. Taverne, P. Merle, M.P. Colombo, A. O'Garra, G. Trinchieri and C. Caux. Reversal of tumor-induced dendritic cell paralysis by CpG immunostimula-tory oligonucleotide and anti-interleukin 10 receptor antibody. *J Exp Med* **196**(4):541–549 (2002).

19. J. Folkman. Role of angiogenesis in tumor growth and metastasis. *Semin Oncol* **29**(6 Suppl 16):15–18 (2002).

20. T. Tonini, F. Rossi and P.P. Claudio. Molecular basis of angiogenesis and cancer. *Oncogene* **22**(42):6549–6556 (2003).

21. D. Ribatti. The crucial role of vascular permeability factor/vascular endothelial growth factor in angiogenesis: a historical review. *Br J Haematol* **128**(3):303–309 (2005).

22. L. Zhang, J.R. Conejo-Garcia, D. Katsaros, P.A. Gimotty, M. Massobrio, G. Reg-nani, A. Makrigiannakis, H. Gray, K. Schlienger, M.N. Liebman, S.C. Rubin and G. Coukos. Intratumoral T cells, recurrence, and survival in epithelial ovarian cancer. *N Engl J Med* **348**(3):203–213 (2003).

23. L. Zhang, N. Yang, J.R. Garcia, A. Mohamed, F. Benencia, S.C. Rubin, D. Allman and G. Coukos. Generation of a syngeneic mouse model to study the effects of vas-cular endothelial growth factor in ovarian carcinoma. *Am J Pathol* **161**(6):2295–2309 (2002).

24. J.R. Conejo-Garcia, R.J. Buckanovich, F. Benencia, M.C. Courreges, S.C. Rubin, R.G. Carroll and G. Coukos. Vascular leukocytes contribute to tumor vasculariza-tion. *Blood* **105**(2):679–681 (2005).

25. J.R. Conejo-Garcia, F. Benencia, M.C. Courreges, E. Kang, A. Mohamed-Hadley, R.J. Buckanovich, D.O. Holtz, A. Jenkins, H. Na, L. Zhang, D.S. Wagner, D. Katsaros, R. Caroll and G. Coukos. Tumor-infiltrating dendritic cell precursors recruited by a beta-defensin contribute to vasculogenesis under the influence of Vegf-A. *Nat Med* **10**(9):950–958 (2004).

26. M.C. Courreges, F. Benencia, J.R. Conejo-Garcia, L. Zhang and G. Coukos. Preparation of apoptotic tumor cells with replication-incompetent HSV augments the efficacy of dendritic cell vaccines. *Cancer Gene Ther* **13**(20):182–193 (2006).

27. D.I. Gabrilovich, H.L. Chen, K.R. Girgis, H.T. Cunningham, G.M. Meny, S. Nadaf, D. Kavanaugh and D.P. Carbone. Production of vascular endothelial growth factor by human tumors inhibits the functional maturation of dendritic cells. *Nat Med* **2**(10):1096–1103 (1996).

28. D.I. Gabrilovich, T. Ishida, S. Nadaf, J.E. Ohm and D.P. Carbone. Antibodies to vascular endothelial growth factor enhance the efficacy of cancer immunotherapy by improving endogenous dendritic cell function. *Clin Cancer Res* **5**(10):2963–2970 (1999).

14

CD4+ T Cells Cooperate with Macrophages for Specific Elimination of MHC Class II-Negative Cancer Cells

Alexandre Corthay

1. INTRODUCTION

Our present knowledge of how T cells eliminate cancer is mainly based on memory immune responses investigated with vaccinated mice[1,2]. In-vivo depletion studies with anti-CD4 monoclonal antibodies (mAb) have revealed that the antitumor immunity conferred by prophylactic vaccination is usually CD4[+] T cell dependent. CD4[+] T cells were required for vaccination-induced immunity against the B16 melanoma[3-5], against the Mc51.9 fibrosarcoma[6], against the J558 plasmacytoma[7], and against the A20 lymphoma[7].

The crucial function of CD4[+] T cells is further supported by the observation that antitumor immunity can be transmitted by adoptive transfer of CD4[+] T cells from vaccinated mice to naive recipients. Transfer of tumor-specific CD4[+] T cells protected mice against challenge with the X5563 and MOPC315 plasmacytomas[8,9], against the 6132A-PRO UV light-induced tumor[10], and against the P815 mastocytoma[11].

Most tumor cells do not express major histocompatibility complex class II (MHC-II) molecules and thus cannot be directly recognized by tumor-specific CD4[+] T cells. Therefore, rejection of MHC-II-negative tumor cells by CD4[+] T cells is most likely dependent on professional antigen-presenting cells (APCs) that endocytose, process, and present tumor antigens on their MHC-II to tumor-specific CD4[+] T cells. This hypothesis is supported by the observation that dendritic cells isolated from large tumors are loaded with tumor antigens and can activate tumor-specific CD4[+] T cells[12,13].

Institute of Immunology, University of Oslo and Rikshospitalet-Radiumhospitalet Medical Center, 0027 Oslo, Norway. Alexandre.Corthay@medisin.uio.no

2. CD4+ T CELLS HELP CD8+ T CELLS TO KILL TUMOR CELLS

The crucial role of tumor-specific CD4[+] T cells in helping cytotoxic CD8[+] T cells kill tumor cells is well-accepted[11,14-16]. Induction of tumor-specific CD4[+] T cells by vaccination with a specific viral MHC-II-restricted epitope resulted in protective immunity against RMA, a virus-induced T lymphoma cell line[14]. In this model, the main effector cells responsible for tumor eradication were identified as tumor-specific CD8[+] cytotoxic T cells[14]. Transfer of tumor-specific Th1 and Th2 CD4[+] T cells protected mice against challenge with the P815 mastocytoma[11]. This protection was abolished by treatment with depleting anti-CD8 mAb[11].

In a more recent study, CD4[+] T cells were shown to be essential for activation of memory CD8[+] T cells to kill tumor cells[16]. Tumor-specific CD4[+] T cells were required for reactivation of memory CD8[+] T cells by tumor antigens presented indirectly. In contrast to memory CD8[+] T cells, effector CD8[+] T cells did not need help from CD4[+] T cells to kill tumor cells[16].

3. CD4+ T CELLS CAN REJECT TUMORS IN THE ABSENCE OF CD8+ T CELLS

Several reports have demonstrated that CD4[+] T cells can reject tumors in the absence of CD8[+] T cells[4,8-10,17]. Transfer of tumor-specific CD4[+] T cells into T cell-depleted or T cell-deficient mice conferred protection against the X5563 and MOPC315 plasmacytomas[8,9]. Similarly, transfer of tumor-specific CD4[+] T cells protected T and B cell-deficient severe combined immunodeficient (SCID) mice against challenge with the 6132A-PRO UV light-induced tumor[10].

It has been proposed that CD4[+] T cells eliminate tumors through activation and recruitment of effector cells, including macrophages and eosinophils[5]. Several studies suggest that cytokines such as interferon-γ (IFNγ) that are secreted by Th1 cells might be involved in antitumor and anti-angiogenic activities[6,10,18].

4. CANCER IMMUNOTHERAPY BY ADOPTIVE TRANSFER OF TUMOR-SPECIFIC CD4+ T CELLS

Adoptive transfer of tumor-specific T cells can efficiently cure mice with cancer[11,15,19-22]. Transfer of tumor-specific CD4[+] T cells protected mice from challenge with the MHC-II-negative Friend virus-induced leukemia FBL-3[17,19]. Tumor eradication was shown not to require participation of cytotoxic CD8[+] T cells[17]. In another model, both ovalbumin-specific Th1 and Th2 cells could cure mice from subcutaneous (s.c.) tumors of the MHC-II-positive A20 lymphoma transduced with ovalbumin[21]. Both Th1 and Th2 cells required CD8[+] T cells to eliminate tumors, and neither of these cells were able to completely eliminate A20– ovalbumin tumors from T and B cell-deficient RAG2-/- mice[21].

Th1 cells have been reported to cure mice with cancer through activation of tumor-specific CD8⁺ T cells[15]. This was shown in a model of immunotherapy of pulmonary nodules of the WP4 fibrosarcoma transduced with β-galactosidase. In this model, the therapeutic effect of transferred β-galactosidase-specific CD4⁺ T cells required the presence of CD8⁺ T cells in the recipient. Transfer of β-galactosidase-specific CD4⁺ T cells was further shown to result in de novo generation of CD8⁺ T cells with specificity to that antigen[15].

5. CD4+ T CELLS IN CANCER IMMUNOSURVEILLANCE

The immune system has been proposed to specifically recognize and eliminate newly transformed cells[23]. A series of reports with gene-targeted mice has recently provided strong experimental support for this cancer immunosurveillance hypothesis[24-30]. A number of lymphocyte subsets (like CD8⁺ T cells, γδ T cells, NK cells, and NKT cells) and effector mechanisms (such as IFNγ, perforin, and TRAIL) have been shown to be critical for immunosurveillance against various types of malignancies[24-30]. These studies in animals support earlier observations that humans with a reduced immune capacity are more prone to develop malignancies[31,32].

The mechanisms of cancer immunosurveillance by CD4⁺ T cells have been recently investigated in a T cell receptor (TCR)-transgenic mouse system[33]. In these transgenic mice, T cells recognize a tumor-specific idiotopic (Id) peptide from the secreted immunoglobulin L-chain V region of the MOPC315 mouse myeloma, presented in the context of MHC-II I-E^d. The high frequency of naive tumor-specific CD4⁺ T cells in TCR-transgenic mice renders the mice resistant to s.c. injection with syngeneic MOPC315 tumor cells, whereas non-transgenic mice develop fatal tumors. Protection is Id-specific, CD4⁺ T cell-mediated, and MHC-II restricted, and does not require the presence of B cells, γδ T cells, and CD8⁺ T cells[9,34,35]. Importantly, MOPC315 lacks MHC-II and therefore cannot be directly recognized by transgenic Id-specific CD4⁺ T cells[12]. Rejection of MOPC315 by Id-specific TCR-transgenic mice does not require immunization of the mice, and thus represents a genuine primary immune response[34].

5.1. Injection of Tumor Cells in Matrigel

It has been difficult to study the mechanisms of tumor rejection in Id-specific TCR-transgenic mice, because the myeloma cells could not be precisely localized in vivo after injection. To solve this, injected tumor cells were embedded in a collagen gel (Matrigel[36]) that is soluble at +4°C but gels at body temperature, resulting in a plug that can easily be identified in vivo (Figure 1)[33]. Another advantage of this technique is that it allows analysis of the infiltrating host cells trapped in the gel. Matrigel is a soluble basement membrane derived from a murine tumor and therefore represents a genuine tumor cell microenvironment[36].

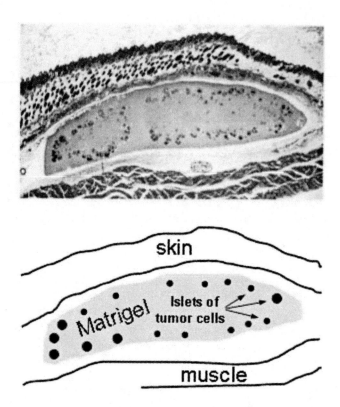

Figure 1. Injection of tumor cells in Matrigel. In order to precisely localize the tumor cells in vivo after subcutaneous injection, injected tumor cells were embedded in a collagen gel (Matrigel). This technique enables analysis of the early interactions between tumor cells and infiltrating cells from the host. Adapted with permission from A. Corthay, D.K. Skovseth, K.U. Lundin, E. Rosjo, H. Omholt, P.O. Hofgaard, G. Haraldsen and B. Bogen. Primary antitumor immune response mediated by CD4+ T cells. *Immunity* **22**:371–383 (2005). Copyright © 2005, Elsevier Inc.

Matrigel constituents (mainly laminin, type IV collagen, and heparan sulfate proteoglycan) are physiological components of the extracellular matrix and thus not expected to hinder migration of leukocytes[36]. TCR-transgenic SCID mice were effectively protected against MOPC315 injected in Matrigel for more than 50 days after injection (Figure 2)[33]. However, most transgenic mice failed to completely reject the myeloma cells injected in Matrigel. As a consequence of this incomplete rejection, slow-growing tumors developed in 3 out of 7 TCR-transgenic SCID mice as late as 60–90 days after injection (Figure 2)[33].

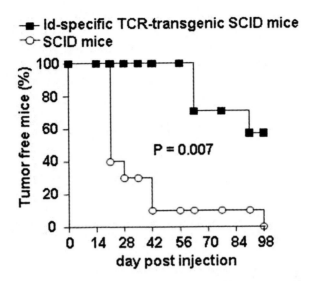

Figure 2. Idiotype (Id)-specific TCR-transgenic SCID mice are protected against syngeneic MOPC315 myeloma cells injected s.c. in a collagen gel (Matrigel). Adapted with permission from A. Corthay, D.K. Skovseth, K.U. Lundin, E. Rosjo, H. Omholt, P.O. Hofgaard, G. Haraldsen and B. Bogen. Primary antitumor immune response mediated by CD4+ T cells. *Immunity* **22**:371–383 (2005). Copyright © 2005, Elsevier Inc.

5.2. Naive Tumor-Specific CD4+ T Cells Become Activated in Draining Lymph Nodes (LN), Migrate to the Incipient Tumor Site and Secrete Cytokines

Proliferation of naive tumor-specific CD4$^+$ T cells was first observed in the LN draining the injection site at day 3 and dramatically increased at day 6 after injection of MOPC315 myeloma cells in Matrigel[33]. This clonal expansion of the tumor-specific CD4$^+$ T cells was associated with upregulation of the activation marker CD69. The immune response was local, as similar changes were not observed in non-draining LN or in the spleen. T cell activation was Id-specific since no response was seen after injection with the control J558 myeloma, which secretes a monoclonal IgA with V-regions different from that of MOPC315. From day 3 to 9 after MOPC315 cell injection, the tumor-specific CD4$^+$ T cells differentiated in draining LN from naive to memory phenotype: the cells increased in size (blast formation), upregulated surface CD11a and CD44 molecules, downregulated CD62L, and synthesized DNA (i.e., incorporated bromodeoxyuridine, BrdU). At day +6, a small but distinct population of CD69$^+$ tumor-specific T cells could be detected inside the Matrigel. At day +9, Matrigel-infiltrating tumor-specific T cells were more frequent and had a typical memory phenotype (enlarged size, CD11a^{hi}, CD44^{hi}, CD62L$^{lo/-}$, BrdU^{+}). Importantly, these Matrigel-infiltrating tumor-specific T cells were producing the Th1 cytokines IFNγ and tumor necrosis factor α at day +11. These results demonstrate that 11 days are sufficient for priming of naive tumor-specific CD4^{+} T cells in draining LN, migration of primed T cells into the tissue where the tumor cells are located, and secretion of cytokines [33].

5.3. Massive Recruitment of Host Macrophages toward the Injected Myeloma Cells

The early in-vivo interactions between the tumor cells and cells of the immune system were visualized by immunostaining of MOPC315-containing Matrigel plugs[33]. At day +6, the nucleated cells inside the Matrigel plug could basically be divided into two distinct populations: (i) The MOPC315 cells growing in islets and stained by the myeloma marker CD138 (syndecan-1), and (ii) the host cells that essentially all expressed MHC-II. These MHC-II^{+} host cells formed a dense layer covering the edge of the myeloma-containing Matrigel plug 3–6 days after injection. Numerous MHC-II^{+} host cells were seen penetrating the Matrigel plug, of which several were in close contact with the CD138^{+} myeloma cells. The Matrigel-infiltrating MHC-II^{+} cells co-expressed the macrophage marker F4/80. These data suggest that there is a massive recruitment of host MHC-II^{+} macrophages toward the tumor cells. These macrophages are most likely derived from blood monocytes that extravasated mainly from vessels situated at the periphery of the plug, but also from vessels surrounded by the gel. Infiltrating T cells could be detected in Matrigel sections, but they were much fewer than the macrophages. Importantly, some tumor-specific T cells apparently made contact with MHC-II^{+} macrophages in the MOPC315-containing Matrigel plug[33].

Flow cytometry was used to characterize the Matrigel-infiltrating MHC-II^{+} cells. Analysis of the cellular content of a Matrigel plug at day +5 revealed a massive infiltration of cells expressing the CD11b (Mac-1) myeloid cell marker. MHC-II expression was almost exclusively restricted to these CD11b^{+} cells. Further characterization of the Matrigel-infiltrating CD11b^{+} cells identified the cells as typical macrophages: they expressed CD11a (LFA-1 α chain), CD54 (ICAM-1), CD80, CD86, and Mac-3, while they were negative for CD45R/B220, CD4 and CD8. Interestingly, a minority (4-12 %) of the Matrigel-infiltrating CD11b^{+} cells expressed CD11c^{+}, indicating a small subset of dendritic cells. Notably, the recruitment of macrophages was dependent on the presence of myeloma cells and not caused by the Matrigel itself since very few macrophages infiltrated cell-free Matrigel plugs[33]. Indeed, already 3 days after injection, myeloma-containing Matrigel plugs were massively infiltrated by macrophages, apparently recruited by chemoattractants secreted by tumor cells[37].

5.4. Tumor-Specific CD4+ T Cells Activate Matrigel-Infiltrating Macrophages

Immunostaining data revealed that some tumor-specific CD4+ T cells made close contact with macrophages inside the MOPC315-containing Matrigel plugs[33]. A functional consequence of such an interaction could be activation of the macrophages by the CD4+ T cells. Indeed, after MOPC315 injections, levels of the activation marker MHC-II on Matrigel-infiltrating macrophages were dramatically increased in Id-specific TCR-transgenic SCID mice as compared to nontransgenic SCID mice. Large numbers of macrophages were found in Matrigel plugs containing MOPC315 as early as 3 days after injection, but at this time point MHC-II expression was not upregulated. In contrast, at day +6 most Matrigel-infiltrating CD11b+ cells had upregulated MHC-II in TCR-transgenic mice, correlating with the influx of tumor-specific CD4+ T cells at the same time point. In order to demonstrate that the observed macrophage activation was mediated by the tumor-specific CD4+ T cells, an anti-CD4 mAb was used to deplete CD4+ T cells in TCR-transgenic SCID mice. The activation of Matrigel-infiltrating macrophages was completely blocked in such CD4+ T cell-depleted mice [33].

To assess the role of antigen presentation in the malignant tissue, an experiment was designed in which tumor-specific CD4+ T cells would be able to meet their cognate antigen in the LN (and be primed) but not in the malignant tissue. For this purpose, MOPC-specific TCR-transgenic SCID mice were injected with MOPC315-containing Matrigel on the right flank and J558-containing Matrigel on the left flank (Figure 3). At day 6 after injection, Matrigel plugs and draining LN were analyzed individually. Several observations could be made from this experiment. First, Id-specific CD4+ T cells became activated in the right flank LN, draining MOPC315-containing Matrigel, but not in the left flank LN, draining J558-containing Matrigel (Figure 3, upper part). This demonstrates that the initial priming of tumor-specific CD4+ T cells is taking place locally, in the LN draining the malignant tissue, rather than systemically. Second, primed Id-specific CD4+ T cells migrated to the same extent into both MOPC315- and J558-containing Matrigel plugs. This reveals that, in contrast to the priming in LN, migration of primed CD4+ T cells into tumor tissues is not antigen-specific. Third, macrophages were recruited to the same extent into MOPC315- and J558-containing Matrigel plugs. Fourth, macrophages were activated in MOPC315-containing Matrigel but not in Matrigel with J558, demonstrating the importance of antigen presentation in situ for T cell-mediated macrophage activation (Figure 3, lower part). Fifth, the levels of CD69 on Matrigel-infiltrating T cells were higher in the MOPC315-containing Matrigel plug as compared to the J558-containing Matrigel plug. This strongly suggests that T cell recognition of tumor-derived peptides presented on MHC-II by macrophages results in reactivation of the tumor-specific CD4+ T cells in situ at the incipient tumor site[33].

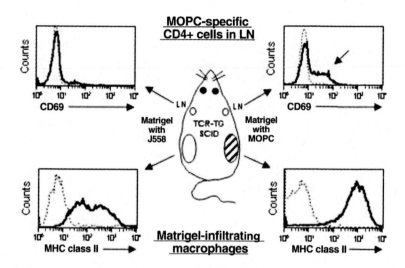

Figure 3. Primary antitumor immune response mediated by CD4⁺ T cells. An MOPC-specific TCR-transgenic SCID mouse was injected s.c. with MOPC315-containing Matrigel on the right flank and J558-containing Matrigel on the left flank. At day +6, Matrigel plugs and draining LN were analyzed individually. Upper part: expression of CD69 on gated CD4⁺ MOPC315-specific T cells in LN draining the left and right flanks. Lower part: expression of MHC-II on gated CD11b⁺ macrophages that had infiltrated the Matrigel plugs containing either J558 (left) or MOPC315 (right). Dotted line: isotype-matched control mAb. Adapted with permission from A. Corthay, D.K. Skovseth, K.U. Lundin, E. Rosjo, H. Omholt, P.O. Hofgaard, G. Haraldsen and B. Bogen. Primary antitumor immune response mediated by CD4+ T cells. *Immunity* **22**:371–383 (2005). Copyright © 2005, Elsevier Inc.

5.5. IFNγ Is Critical for T Cell-Mediated Macrophage Activation and Tumor Rejection

Matrigel-infiltrating tumor-specific CD4⁺ T cells produce IFNγ that is a potent macrophage-activating factor[38]. To test whether the observed activation of Matrigel-infiltrating macrophages was dependent on IFNγ, TCR-transgenic SCID mice were injected with a blocking anti-IFNγ mAb. T cell-mediated activation of Matrigel-infiltrating macrophages was completely inhibited by blocking IFNγ. Notably, macrophage activation could be restored in non-transgenic SCID mice by s.c. injection with mouse recombinant IFNγ. Thus, IFNγ is critical for activation of the macrophages that infiltrate the MOPC315-containing Matrigel plugs[33].

Macrophage activation by IFNγ is referred to as the classical activation pathway and is characterized by upregulation of surface FcγRII/III. A number of alternative pathways for macrophage activation have been described, which are associated with upregulation of receptors like FcεRII (CD23), scavenger RI (CD204), and mannose R (CD206)[39]. Flow cytometry analysis confirmed that Matrigel-infiltrating macrophages in TCR-transgenic SCID mice have a classical activation phenotype (FcγRII/III[hi], CD23[-], CD204[lo/-], CD206 [lo/-]), in accordance with their IFNγ-mediated activation[33].

The importance of IFNγ in tumor rejection by CD4[+] T cells was investigated. In a first experiment, blockage of IFNγ rendered the TCR-transgenic SCID mice susceptible to MOPC315 tumor development, revealing that the antitumor response was dependent on IFNγ. In a second complementary experiment, s.c. injection with mouse recombinant IFNγ was shown to significantly delay MOPC315 tumor growth in SCID mice. This indicates that the antitumor function of tumor-specific CD4[+] T cells can at least partly be substituted with local injection of IFNγ[33].

5.6. T Cell-Activated Macrophages Suppress Tumor Cell Growth

The primary anti-MOPC315 immune response is associated with secretion of IFNγ by tumor-specific CD4[+] T cells, resulting in activation of macrophages in close proximity to the tumor cells[33]. This suggests that IFNγ- and T cell-activated macrophages could directly exert tumor suppressive functions. Addition of IFNγ in high concentration to MOPC315 cells in vitro had no effect on cell proliferation, demonstrating that IFNγ per se was not directly cytotoxic to the myeloma cells. Similarly, Matrigel-infiltrating macrophages isolated from control SCID mice had no influence on MOPC315 in-vitro growth (Figure 4). In striking contrast, Matrigel-infiltrating macrophages isolated from TCR-transgenic SCID mice at day +7 were able to completely inhibit the in-vitro proliferation of MOPC315 cells in a dose-dependent manner (Figure 4). These results demonstrate that Matrigel-infiltrating macrophages that have been activated in vivo by tumor-specific CD4[+] T cells can effectively inhibit the growth of MOPC315 cells in vitro. Since the suppression of tumor cell growth was observed in a short-term (2.5 days) ex-vivo assay, it is likely that the activated macrophages exert the same inhibitory function in vivo[33].

It is well established that solid tumors are infiltrated by macrophages, but the function of these tumor-associated macrophages (TAMs) is controversial. Initial studies revealed that peritoneal macrophages from immunized mice could kill tumor cells in vitro, and the reaction was shown to be immunologically specific[40,41]. Subsequent in-vitro studies demonstrated that macrophages could be rendered tumoricidal by treatment with IFNγ[38]. However, extensive infiltration of tumors by macrophages is often associated with a poor prognosis[42]. Therefore, it was proposed that the tumor microenvironment may educate TAMs, so that they produce important growth factors and enzymes that stimulate angiogenesis and tumor growth[39,42,43]. More specifically, tumor cells have been suggested to redirect TAM activity by secreting cytokines like IL-10 and TGF-β. Thus,

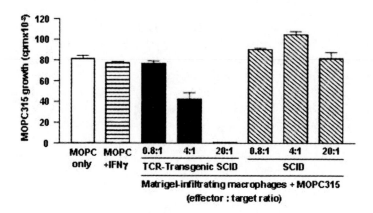

Figure 4. T cell-activated macrophages suppress tumor cell growth. MOPC315-specific TCR-transgenic SCID or SCID mice were injected with Matrigel containing MOPC315. Matrigel-infiltrating CD11b⁺ macrophages were purified at day +7 and tested at various effector to target ratios for their ability to suppress the proliferation of MOPC315 cells in vitro in a growth inhibition assay. In the same experiment, the direct cytotoxicity of IFNγ (5000 U/ml) on MOPC315 was tested. Adapted with permission from A. Corthay, D.K. Skovseth, K.U. Lundin, E. Rosjo, H. Omholt, P.O. Hofgaard, G. Haraldsen and B. Bogen. Primary antitumor immune response mediated by CD4+ T cells. *Immunity* **22**:371–383 (2005). Copyright © 2005, Elsevier Inc.

TAMs may have either tumor-suppressing or tumor-promoting functions, depending on their activation state[44,45]. Tumor-specific CD4⁺ T cells may have a pivotal role in preventing early tumorigenesis by secreting IFNγ and inducing the classical macrophage activation pathway, which results in inhibition of tumor cell growth[33]. A proposed mechanism for immunosurveillance of MHC-II-negative cancer cells by tumor-specific CD4⁺ T cells through collaboration with macrophages is summarized in Figure 5.

6. CONCLUSIONS

CD4⁺ T helper cells have the well-known function of regulating the adaptive immune response against pathogens. Surprisingly, the role of CD4⁺ T cells in tumor immunity has been largely neglected, while extensive research was conducted on tumor-specific CD8⁺ T cells. The role of CD4⁺ T cells in fighting cancer has been conceptually problematic because most tumor cells do not express MHC class II (MHC-II) and thus cannot be directly recognized by tumor-specific CD4⁺ T cells. Recent data reveal that specific recognition and elimination of MHC-II-negative tumor cells can be efficiently achieved through collaboration between tumor-specific CD4⁺ T cells and macrophages. MHC-II-

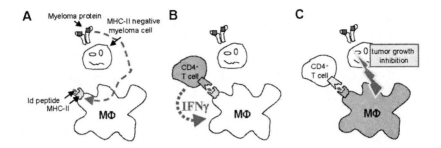

Figure 5. Mechanism for immunosurveillance of MHC-II-negative cancer cells by tumor-specific CD4⁺ T cells. (A) MHC-II-negative myeloma cells injected s.c. into syngeneic mice are surrounded within 3 days by macrophages that capture tumor antigens. (B) Within 6 days, naive myeloma-specific CD4⁺ T cells become activated in draining LN and subsequently migrate to the incipient tumor site. After recognition of tumor-derived peptides presented on MHC-II by macrophages, the tumor-specific CD4⁺ T cells are reactivated and start to secrete cytokines. T cell-derived IFNγ activate macrophages in close proximity to tumor cells. (C) Tumor cell growth is inhibited by such locally activated macrophages.

negative myeloma cells injected subcutaneously into syngeneic mice were surrounded within 3 days by macrophages that captured tumor antigens. Within 6 days, naive myeloma-specific CD4⁺ T cells became activated in draining LN and subsequently migrated to the incipient tumor site. Upon recognition of tumor-derived antigenic peptides presented on MHC-II by macrophages, the myeloma-specific CD4⁺ T cells were reactivated and started to secrete cytokines. T cell-derived IFNγ activated macrophages in close proximity to the tumor cells. Tumor cell growth was completely inhibited by such locally activated macrophages.

7. ACKNOWLEDGMENTS

This work was supported by grants from the Research Council of Norway, the Multiple Myeloma Research Foundation, the Norwegian Cancer Society, and Medinnova.

8. REFERENCES

1. L. Gross. Intradermal immunization of C3H mice against a sarcoma that originated in an animal of the same line. *Cancer Res* **3**:326–333 (1943).
2. R.G. Lynch, R.J. Graff, S. Sirisinha, E.S. Simms and H.N. Eisen. Myeloma proteins as tumor-specific transplantation antigens. *Proc Natl Acad Sci USA* **69**:1540–1544 (1972).

3. G. Dranoff, E. Jaffee, A. Lazenby, P. Golumbek, H. Levitsky, K. Brose, V. Jackson, H. Hamada, D. Pardoll and R.C. Mulligan. Vaccination with irradiated tumor cells engineered to secrete murine granulocyte-macrophage colony-stimulating factor stimulates potent, specific, and long-lasting anti-tumor immunity. *Proc Natl Acad Sci USA* **90**:3539–3543 (1993).

4. H.I. Levitsky, A. Lazenby, R.J. Hayashi and D.M. Pardoll. In vivo priming of two distinct antitumor effector populations: the role of MHC class I expression. *J Exp Med* **179**:1215–1224 (1994).

5. K. Hung, R. Hayashi, A. Lafond-Walker, C. Lowenstein, D. Pardoll and H. Levitsky. The central role of CD4(+) T cells in the antitumor immune response. *J Exp Med* **188**:2357–2368 (1998).

6. Z. Qin and T. Blankenstei. CD4+ T cell-mediated tumor rejection involves inhibition of angiogenesis that is dependent on IFN gamma receptor expression by non-hematopoietic cells. *Immunity* **12**:677–686 (2000).

7. K. Liu, J. Idoyaga, A. Charalambous, S. Fujii, A. Bonito, J. Mordoh, R. Wainstok, X.F. Bai, Y. Liu and R.M. Steinman. Innate NKT lymphocytes confer superior adaptive immunity via tumor-capturing dendritic cells. *J Exp Med* **202**:1507–1516 (2005).

8. H. Fujiwara, M. Fukuzawa, T. Yoshioka, H. Nakajima and T. Hamaoka. The role of tumor-specific Lyt-1+2- T cells in eradicating tumor cells in vivo, I: lyt-1+2- T cells do not necessarily require recruitment of host's cytotoxic T cell precursors for implementation of in vivo immunity. *J Immunol* **133**:1671–1676 (1984).

9. B. Bogen, L. Munthe, A. Sollien, P. Hofgaard, H. Omholt, F. Dagnaes, Z. Dembic and G.F. Lauritzsen. Naive CD4+ T cells confer idiotype-specific tumor resistance in the absence of antibodies. *Eur J Immunol* **25**:3079–3086 (1995).

10. D. Mumberg, P.A. Monach, S. Wanderling, M. Philip, A.Y. Toledano, R.D. Schreiber and H. Schreiber. CD4(+) T cells eliminate MHC class II-negative cancer cells in vivo by indirect effects of IFN-gamma. *Proc Natl Acad Sci USA* **96**:8633–8638 (1999).

11. F. Fallarino, U. Grohmann, R. Bianchi, C. Vacca, M.C. Fioretti and P. Puccetti. Th1 and Th2 cell clones to a poorly immunogenic tumor antigen initiate CD8+ T cell-dependent tumor eradication in vivo. *J Immunol* **165**:5495–5501 (2000).

12. Z. Dembic, K. Schenck and B. Bogen. Dendritic cells purified from myeloma are primed with tumor-specific antigen (idiotype) and activate CD4+ T cells. *Proc Natl Acad Sci USA* **97**:2697–2702 (2000).

13. Z. Dembic, J.A. Rottingen, J. Dellacasagrande, K. Schenck and B. Bogen. Phagocytic dendritic cells from myelomas activate tumor-specific T cells at a single cell level. *Blood* **97**:2808–2814 (2001).

14. F. Ossendorp, E. Mengede, M. Camps, R. Filius and C.J. Melief. Specific T helper cell requirement for optimal induction of cytotoxic T lymphocytes against major histocompatibility complex class II negative tumors. *J Exp Med* **187**:693–702 (1998).

15. D.R. Surman, M.E. Dudley, W.W. Overwijk and N.P. Restifo. Cutting edge: CD4+ T cell control of CD8+ T cell reactivity to a model tumor antigen. *J Immunol* **164**:562–565 (2000).

16. F.G. Gao, V. Khammanivong, W.J. Liu, G.R. Leggatt, I.H. Frazer and G.J. Fernando. Antigen-specific CD4+ T-cell help is required to activate a memory CD8+ T cell to a fully functional tumor killer cell. *Cancer Res* **62**:6438–6441 (2002).

17. P.D. Greenberg, D.E. Kern and M.A. Cheever. Therapy of disseminated murine leukemia with cyclophosphamide and immune Lyt-1+,2- T cells: tumor eradication does not require participation of cytotoxic T cells. *J Exp Med* **161**:1122–1134 (1985).

18. C.M. Coughlin, K.E. Salhany, M.S. Gee, D.C. LaTemple, S. Kotenko, X. Ma, G. Gri, M. Wysocka, J.E. Kim, L. Liu, F. Liao, J.M. Farber, S. Pestka, G. Trinchieri and W.M. Lee. Tumor cell responses to IFNgamma affect tumorigenicity and response to IL-12 therapy and antiangiogenesis. *Immunity* **9**:25–34 (1998).

19. P.D. Greenberg, M.A. Cheever and A. Fefer. Eradication of disseminated murine leukemia by chemoimmunotherapy with cyclophosphamide and adoptively transferred immune syngeneic Lyt-1+2- lymphocytes. *J Exp Med* **154**:952–963 (1981).

20. M. Kahn, H. Sugawara, P. McGowan, K. Okuno, S. Nagoya, K.E. Hellstrom, I. Hellstrom and P. Greenberg. CD4+ T cell clones specific for the human p97 melanoma-associated antigen can eradicate pulmonary metastases from a murine tumor expressing the p97 antigen. *J Immunol* **146**:3235–3241 (1991).

21. T. Nishimura, K. Iwakabe, M. Sekimoto, Y. Ohmi, T. Yahata, M. Nakui, T. Sato, S. Habu, H. Tashiro, M. Sato and A. Ohta. Distinct role of antigen-specific T helper type 1 (Th1) and Th2 cells in tumor eradication in vivo. *J Exp Med* **190**:617–627 (1999).

22. K.U. Lundin, P.O. Hofgaard, H. Omholt, L.A. Munthe, A. Corthay and B. Bogen. Therapeutic effect of idiotype-specific CD4+ T cells against B-cell lymphoma in the absence of anti-idiotypic antibodies. *Blood* **102**:605–612 (2003).

23. F.M. Burnet. The concept of immunological surveillance. *Prog Exp Tumor Res* **13**:1–27 (1970).

24. D.H. Kaplan, V. Shankaran, A.S. Dighe, E. Stockert, M. Aguet, L.J. Old and R.D. Schreiber. Demonstration of an interferon gamma-dependent tumor surveillance system in immunocompetent mice. *Proc Natl Acad Sci USA* **95**:7556–7561 (1998).

25. M.J. Smyth, K.Y. Thia, S.E. Street, E. Cretney, J.A. Trapani, M. Taniguchi, T. Kawano, S.B. Pelikan, N.Y. Crowe and D.I. Godfrey. Differential tumor surveillance by natural killer (NK) and NKT cells. *J Exp Med* **191**:661–668 (2000).

26. M.J. Smyth, K.Y. Thia, S.E. Street, D. MacGregor, D.I. Godfrey and J.A. Trapani. Perforin-mediated cytotoxicity is critical for surveillance of spontaneous lymphoma. *J Exp Med* **192**:755–760 (2000).

27. V. Shankaran, H. Ikeda, A.T. Bruce, J.M. White, P.E. Swanson, L.J. Old and R.D. Schreiber. IFNgamma and lymphocytes prevent primary tumour development and shape tumour immunogenicity. *Nature* **410**:1107–1111 (2001).

28. M. Girardi, D.E. Oppenheim, C.R. Steele, J.M. Lewis, E. Glusac, R. Filler, P. Hobby, B. Sutton, R.E. Tigelaar and A.C. Hayday. Regulation of cutaneous malignancy by gammadelta T cells. *Science* **294**:605–609 (2001).

29. S.E. Street, J.A. Trapani, D. MacGregor and M.J. Smyth. Suppression of lymphoma and epithelial malignancies effected by interferon gamma. *J Exp Med* **196**:129–134 (2002).

30. K. Takeda, M.J. Smyth, E. Cretney, Y. Hayakawa, N. Kayagaki, H. Yagita and K. Okumura. Critical role for tumor necrosis factor-related apoptosis-inducing ligand in immune surveillance against tumor development. *J Exp Med* **195**:161–169 (2002).

31. R.A. Gatti and R.A. Good. Occurrence of malignancy in immunodeficiency diseases: a literature review. *Cancer* **28**:89–98 (1971).

32. S.A. Birkeland, H.H. Storm, L.U. Lamm, L. Barlow, I. Blohme, B. Forsberg, B. Eklund, O. Fjeldborg, M. Friedberg and L. Frodin. Cancer risk after renal transplantation in the Nordic countries, 1964–1986. *Int J Cancer* **60**:183–189 (1995).

33. A. Corthay, D.K. Skovseth, K.U. Lundin, E. Rosjo, H. Omholt, P.O. Hofgaard, G. Haraldsen and B. Bogen. Primary antitumor immune response mediated by CD4+ T cells. *Immunity* **22**:371–383 (2005).

34. G.F. Lauritzsen, S. Weiss, Z. Dembic and B. Bogen. Naive idiotype-specific CD4+ T cells and immunosurveillance of B-cell tumors. *Proc Natl Acad Sci USA* **91**:5700–5704 (1994).

35. Z. Dembic, P.O. Hofgaard, H. Omholt and B. Bogen. Anti-class II antibodies, but not cytotoxic T-lymphocyte antigen 4-immunoglobulin hybrid molecules, prevent rejection of major histocompatibility complex class II-negative myeloma in T-cell receptor-transgenic mice. *Scand J Immunol* **60**:143–152 (2004).

36. H.K. Kleinman, M.L. McGarvey, J.R. Hassell, V.L. Star, F.B. Cannon, G.W. Laurie and G.R. Martin. Basement membrane complexes with biological activity. *Biochemistry* **25**:312–318 (1986).

37. B. Bottazzi, N. Polentarutti, R. Acero, A. Balsari, D. Boraschi, P. Ghezzi, M. Salmona and A. Mantovani. Regulation of the macrophage content of neoplasms by chemoattractants. *Science* **220**:210–212 (1983).

38. R.D. Schreiber, J.L. Pace, S.W. Russell, A. Altman and D.H. Katz. Macrophage-activating factor produced by a T cell hybridoma: physiochemical and biosynthetic resemblance to gamma-interferon. *J Immunol* **131**:826–832 (1983).

39. A. Mantovani, et al. A. Sica, S. Sozzani, P. Allavena, A. Vecchi and M. Locati. The chemokine system in diverse forms of macrophage activation and polarization. *Trends Immunol* **25**:677–686 (2004).

40. R. Evans and P. Alexander. Cooperation of immune lymphoid cells with macrophages in tumour immunity. *Nature* **228**:620–622 (1970).

41. R. Evans and P. Alexander. Mechanism of immunologically specific killing of tumour cells by macrophages. *Nature* **236**:168–170 (1972).

42. L. Bingle, N.J. Brown and C.E. Lewis. The role of tumour-associated macrophages in tumour progression: implications for new anticancer therapies. *J Pathol* **196**:254–265 (2002).

43. J.W. Pollard. Tumour-educated macrophages promote tumour progression and metastasis. *Nat Rev Cancer* **4**:71–78 (2004).

44. K. Tsung, J.P. Dolan, Y.L. Tsung and J.A. Norton. Macrophages as effector cells in interleukin 12-induced T cell-dependent tumor rejection. *Cancer Res* **62**:5069–5075 (2002).

45. C. Guiducci, A.P. Vicari, S. Sangaletti, G. Trinchieri and M.P. Colombo. Redirecting in vivo elicited tumor infiltrating macrophages and dendritic cells towards tumor rejection. *Cancer Res* **65**:3437–3446 (2005).

15

RECEPTORS AND PATHWAYS IN INNATE ANTIFUNGAL IMMUNITY:

THE IMPLICATION FOR TOLERANCE AND IMMUNITY TO FUNGI

Teresa Zelante, Claudia Montagnoli, Silvia Bozza, Roberta Gaziano, Silvia Bellocchio, Pierluigi Bonifazi, Silvia Moretti, Francesca Fallarino, Paolo Puccetti, and Luigina Romani

1. INTRODUCTION

In the last years, the clinical relevance of fungal diseases has gained importance because of an increasing population of immunocompromised hosts, such as patients who have undergone transplants, patients with various types of leukemia, and people infected with HIV. Although some virulence factors are of obvious importance, pathogenicity cannot be considered an inherent characteristic of fungi.[1] Fungi seem to have a complex relationship with the vertebrate immune system, mainly due to some prominent features: among these, the ability of dimorphic fungi to exist in different forms and to reversibly switch from one to the other in infection. Although association between morphogenesis and virulence has long been presumed for fungi that are human pathogens[2], no molecular data unambiguously establish a role for fungal morphogenesis as a virulence factor. What fungal morphogenesis implicates through antigenic variability, phenotypic switching, and dimorphic transition is the existence of a multitude of recognition and effector mechanisms to oppose fungal infectivity at the different body sites.

Teresa Zelante, Department of Experimental Medicine and Biochemical Sciences, University of Perugia, Perugia, Italy. Address correspondence to: Teresa Zelante, Dept. of Experimental Medicine and Biochemical Sciences, University of Perugia, 06122 Perugia, Italy. Phone: 039.075.585.7498. metzelante@hotmail.com

Most fungi need a stable host–parasite interaction characterized by an immune response strong enough to allow host survival without pathogen elimination, thereby establishing commensalisms. Therefore, the balance of proinflammatory and antiinflammatory signaling is a prerequisite for successful host–fungus interaction. In light of these considerations, although developments in fungal genomics may provide new insights in mechanisms of pathogenicity[3], the responsibility for virulence, regardless the mode of its generation and maintenance, is shared by the host and the fungus at the pathogen–host interface. Studies with *Candida albicans* have provided a paradigm that incorporates contributions from both the fungus and the host to explain the theme of the origin and maintenance of virulence for commensals. Through a high degree of flexibility, the model accommodates the concept of virulence as an important component of fungus fitness in vivo within the plasticity of the host immune system[4].

2. WHAT AND WHICH ARE OPPORTUNISTIC FUNGAL PATHOGENS?

The human commensal *C. albicans* is the leading fungal cause of important diseases in humans[5]. *Candida* is a polymorphic fungus; it can exist in different forms that have distinct shapes: yeast cells, pseudohyphal cells, and true hyphal cells, all of which can be found in infected tissues. The ability to switch from yeast to filamentous form is required for virulence, but much of the evidence linking transition and virulence remains equivocal. The clinical spectrum of *C. albicans* infections ranges from mucocutaneous to systemic life-threatening infections. The predisposing factors to severe candidal infections can be congenital or acquired and concern defects of cell-mediated immunity, including defects in neutrophils and dysregulated Th cell reactivity.

Most fungal infections are of exogenous origin. A striking example is *Aspergillus fumigatus*, a saprophitic and ubiquitous fungus, for the ease of dispersion of its conidia. The diseases caused by *Aspergillus* range from benign colonization and allergy to deadly diseases such as invasive pulmonary aspergillosis.[6] The small conidia can remain in suspension in the environment for a long period of time, reaching human pulmonary alveoli and constantly exposing individuals inhaling them. In immunocompromised hosts, such as neutropenic patients or transplanted patients undergoing graft-versus-host disease, the inhalation can provoke serious diseases, consisting of host tissue invasion by conidia germinated to septate hyphae, an invasive form associated with fatal infections.

3. THE IMMUNE RESPONSE TO FUNGI : FROM MICROBE SENSING TO HOST DEFENCING

Protective immunity against fungal pathogens is achieved by integration of two distinct arms of the immune system — the innate and adaptive (or antigen-specific) responses[7]. The majority of fungi are detected and destroyed within hours by innate defense mechanisms. The innate mechanisms appeared early in the evolution of multicellular organisms and act early after infection. Innate defense strategies are designed to detect broad and conserved patterns that differ between pathogenic organisms and their multicellular hosts. Most of the innate mechanisms are inducible upon infection, and their activation requires specific recognition of invariant evolutionarily conserved molecular structures shared by large groups of pathogens by a set of pattern recognition receptors (PRRs), including Toll-like receptors (TLRs)[8]. In vertebrates, however, if the infectious organism can breach these early lines of defense, an adaptive immune response will ensue, with generation of antigen-specific T helper (Th) effector and B cells that specifically target the pathogen and memory cells that prevent subsequent infection with the same microorganism. Cytokines and other mediators play an essential role in the process and, indeed, may ultimately determine the type of inflammatory response that is generated toward the pathogens. The dichotomous Th-cell model has proven to be a useful construct that sheds light on the general principle that diverse effector functions are required for eradication of different fungal infections[9]. To limit the pathologic consequences of an excessive inflammatory cell-mediated immune reaction, the immune system resorts to a number of protective mechanisms, including the reciprocal crossregulatory effects of Th1- and Th2-type effector cytokines, such as interferon (IFN)-γ and interleukin (IL)-4, and the generation of regulatory T cells (Treg). Thus, innate and adaptive immune responses are intimately linked and controlled by sets of molecules and receptors that act to generate the most effective form of immunity for protection against fungal pathogens (Figure 1).

4. SENSING FUNGI

Innate mechanisms of defense are traditionally divided into constitutive and inducible. Constitutive mechanisms are present at sites of continuous interaction with fungi and include the barrier function of the body surface and the mucosal epithelial surfaces of the respiratory, gastrointestinal, and genitourinary tracts. Additional mechanisms are microbial antagonism, defensins, and collectins. These molecules lead to microbe opsonization and recruitment of phagocytic cells. Innate inducible mechanisms are activated upon specific recognition of microbes through various PRRs. PRRs are expressed on a variety of innate cells, such as epithelial cells, polymorphonuclear cells (PMNs), monocytes, and dendritic cells (DCs). PRRs for fungi include TLRs, receptors for complement components (CRs), for the FC portion of immunoglobulins (FcRs), receptors for mannosyl/fucosyl glycoconjugate ligands (MRs) and for β–glucan (dectin-1) (Figure 1)[10]. Each receptor on phagocytes not only mediates distinct downstream intracellular events related to clearance, but also participates in complex and disparate functions related to immunomodulation and activation of immunity,

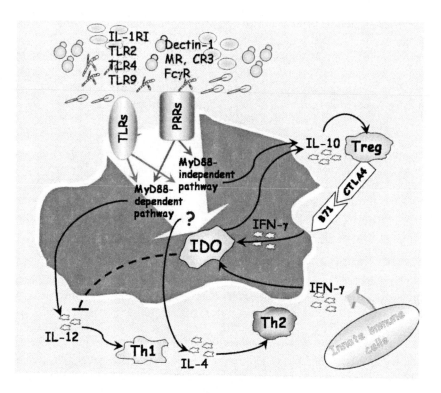

Figure 1. The interface between innate and adaptive immunity to fungi. Essential to the successful removal of fungal pathogens is the early recognition of fungi by components of the innate immune system. These involve the complement system, opsonins, antibodies, IDO-dependent metabolic pathways, and specialized receptors such as TLRs expressed on phagocytes and dendritic cells that recognize specific fungal-derived molecular structures. Successful engagement of some of these pathways leads to an inflammatory response with destruction of the pathogen alongside the establishment of dendritic cell and T cell interactions. A well-orchestrated innate and adaptive immune response will lead to pathogen control and host protective Th1/Treg immunity. Failure to do so can lead to pathogen proliferation and ultimately to dysregulated Th2 immunity. Solid and dotted lines, positive and negative signals.

depending on cell types. The different receptors are called upon to serve as early-warning systems. Not surprisingly, their ability to activate, in isolation, various effector functions is limited. With few exceptions, internalization via constitutively competent MRs does not represent an effective way of clearing fungi in the absence of opsonins. However, MRs on DCs activate specific programs that are relevant for the development of antifungal immune responses (see below). CR3 (also known as CD11b/CD18) engagement is one of the most efficient uptake mechanisms of opsonized fungi, but it has the remarkable characteristic of a broad capacity for recognition of diverse fungal ligands. The multiplicity of binding sites and the existence of different activation states enable

CR3 of disparate (both positive and negative) effector activities against fungi. Thus, because signaling through CR3 may not lead to phagocyte activation without concomitant engagement of FcR, the use of this receptor may contribute to intracellular fungal parasitism. It is of interest, indeed, that *H. capsulatum* uses this receptor for entry into macrophages where it survives[11], but not into DCs, where it is rapidly degraded[12]. Likewise, *Candida* exploits entry through CR3 to survive inside DCs[4]. In contrast, ligation of FcR is usually sufficient to trigger phagocytosis, a vigorous oxidative burst, and generation of proinflammatory cytokines. TLRs, which are broadly distributed on cells of the immune system, are arguably the best-studied immune sensors of invading pathogens, and the signaling pathways that are triggered by pathogen detection initiate innate immunity and help to strengthen adaptive immunity. TLRs belong to the TIR (Toll/interleukin-1 (IL-1) receptor) superfamily, which is divided into two main subgroups: the IL-1 receptors and the TLRs. The IL-1 receptor subgroup consists of at least ten receptors, whose most notable members include both type I and type II receptors, the IL-18 receptor, and the orphan receptors ST2 and single immunoglobulin IL-1-related receptor (SIGIRR; also known as TIR8). The TLR subgroup comprises TLR1–TLR11. All members of this superfamily signal in a similar manner owing to the presence of a conserved TIR domain in the cytosolic region, which activates common signaling pathways, most notably those leading to the activation of the transcription factor nuclear factor-κB (NF-κB) and stress-activated protein kinases. TLR activation is a double-edged sword. It is essential for provoking the innate response and enhancing adaptive immunity against pathogens[13]. However, members of the TLR family are also involved in the pathogenesis of autoimmune, chronic inflammatory, and infectious diseases — by hyperinduction of proinflammatory cytokines, by facilitating tissue damage, or by impaired protective immunity.

The different impact of TLRs on the occurrence of the innate and adaptive Th immunity to fungi is consistent with the ability of each individual TLR to activate specialized antifungal effector functions on phagocytes and DCs[14,15]. Although not affecting the phagocytosis, TLRs affect specific antifungal programs of phagocytes, such as the respiratory burst, and degranulation and production of chemokines and cytokines (see below). As the quantity and specificity of delivery of toxic neutrophil products ultimately determine the relative efficiency of fungicidal activity versus inflammatory cytotoxicity to host cells, this implicates that TLRs may contribute to protection and immunopathology against fungi[13]. The emerging picture calls for: (i) the essential requirement for the IL-1RI/MyD88-dependent pathway in the innate and Th1-mediated resistance to *C. albicans*; (ii) the essential requirement of the MyD88-dependent pathway in recognition of and response to *C. neoformans*; (iii) the crucial involvement, although not essential, of the TLR4/MyD88 pathway in recognition of and resistance to *A. fumigatus;* (iv) the beneficial effect of TLR9 stimulation on immune-mediated resistance to pulmonary aspergillosis and cryptococcosis; (v) the dependency of TLR-dependent pathways from the site of the infection, also illustrated in flies infected with *C. neoformans*, and (iv) the occurrence of TLR signaling in a morphotype-specific manner, although the simultaneous engagement of multiple TLRs as well as TLR cooperativity in vivo makes it difficult to gauge the relative contribution of each single fungal morphotype in TLR activation and functioning[16,17]. For instance, TLR4 and CD14 mediate the recognition of *C. albicans*-derived mannan. However, TNF-α and IL-1β production in response to *Candida* may also occur in a TLR4-independent manner, a finding consistent with the observation that resistance to infection is decreased in TLR4-deficient mice along with release of chemokines. Therefore, TLR2 and TLR4 are both implicated in the elicitation of host defense to the fungus, a finding exemplifying recruitment of different TLRs by one microbial species.

5. TUNING THE ADAPTIVE IMMUNE RESPONSES: THE INSTRUCTIVE ROLE OF DCs

Since their original discovery in 1973, DCs have assumed center stage as key players in the initiation of adaptive immunity. They consist of a family of different subpopulations, which originate from the bone marrow and develop to perform different functions. In infections they are central in the regulation of balance between immunopathology and protective immunity generated by host–microbe interactions. At the immature stages of development, DCs capture antigen-derived information in peripheral tissues, acting as *sentinel cells*. They are located at the epithelial barriers that often represent the major portals of pathogen entry, take up antigens avidly, and move into secondary lymphoid organs activating both helper and cytotoxic T cells. DC maturation is characterized by downregulation of antigen acquisition, increased expression of major histocompatibility complex (MHC), costimulatory molecules, IL-12 production, and altered expression of chemokine receptors. In addition to initiating immunity, DCs can also downregulate immune responses, resulting in induction of peripheral tolerance and prevention of autoimmune disorders[18].

It is now clear that DCs, through different receptors, are able to recognize and capture fungus-associated information, translating it in qualitatively different Th immune responses, in vitro and in vivo[7]. Human and murine DCs internalize many different fungi such as *A. fumigatus*, *C. albicans*, *Cryptococcus neoformans*, *Hystoplasma capsulatum*, *Coccidioides immitis*, and *Malassezia furfur*. *A. fumigatus* and *C. albicans* have been successfully used as pathogen models to dissect events occurring at the fungus/DC interface. DCs internalize *Candida* yeasts, *Aspergillus* conidia, and the hyphae of both. Phagocytosis occurs in different ways and involves different PRRs. Recognition and internalization of yeasts and conidia occur mainly by coiling phagocytosis, through the ligation of MRs of different sugar specificity, DC-SIGN and, partly, CR3. The entry of hyphae occurs by a more conventional, zipper-type phagocytosis, and involves the cooperative action of FcγR and CR3. Phagocytosis does not require TLR2, TLR4, TLR9, and MyD88.

The engagement of distinct receptors by different fungal morphotypes translates into downstream signaling events, ultimately regulating cytokine production and costimulation, an event greatly influenced by fungal opsonins. Entry of *Candida* yeasts or *Aspergillus* conidia through MRs results in the production of proinflammatory cytokines, including IL-12, upregulation of costimulatory molecules and histocompatibility Class II antigens, and activation of protective Th1 cell responses. IL-12 production by DCs also occurs through the MyD88 pathway with the implication of distinct TLRs (IL-1RI and TLR9 for *Candida* and TLR4 and TLR9 for *Aspergillus*). Actually, these events are all suppressed upon entry through CR3. In contrast, coligation of CR3 with FcγR, as in the phagocytosis of hyphae, results in the production of IL-4/IL-10, upregulation of costimulatory molecules and histocompatibility Class II antigens, and activation of Th2/Treg cells. Opsonization with MBL, C3, and/or IgG greatly modifies receptor exploitation on DCs by the different fungal morphotypes and qualitatively affects DCs activation. Thus, collectins appear to favor the phagocytosis of the fungus without implicating the production of cytokine messengers to the immune system, an activity compatible with a primitive mechanism of host defense and in line with their ability to downregulate the inflammatory response to fungi. All together, opsonins, by subverting the morphotype-specific program of activation of DCs, may qualitatively affect DC and Th functioning.

Studies in vivo confirm that DCs sample fungi at sites of infection, transport them to the draining lymph nodes, and initiate disparate Th responses to the different fungal morphotypes[19,20]. Furthermore, adoptive transfer of ex-vivo DCs transfected with fungal RNA restores protective antifungal immunity in a murine model of allogeneic bone marrow transplantation[19,21]. These results, along with the finding that fungus-pulsed DCs could reverse the T cell anergy of patients with fungal diseases, may suggest the utility of DCs for fungal vaccines.

6. DCs AS TOLERANCE MEDIATORS VIA TRYPTOPHAN CATABOLISM

The inflammatory reaction, although useful for the microbicidal functions, may also contribute to pathogenicity. Recovery from infection may not only depend on fungal growth restriction but also on resolution of inflammatory pathology. This resolution imposes a new job on the immune system. In addition to efficient control of pathogens, tight regulatory mechanisms are required in order to balance protective immunity and immunopathology. Experimental and clinical evidence indicates that the course and outcome of infections not only correlate with fungal load but also with immunopathology[1]. Recent evidence suggests that the inflammatory/antiinflammatory state of DCs in response to fungi is strictly controlled by the metabolic pathway involved in tryptophan catabolism and mediated by the enzyme indoleamine 2,3-dioxygenase (IDO)[22]. IDO has a complex role in immunoregulation in infection, pregnancy, autoimmunity, transplantation, and neoplasia. IDO expressing DCs are regarded as regulatory DCs specialized to cause antigen-specific deletional tolerance or otherwise negatively regulating responding T cells. IFN-γ is required for functional IDO enzymatic activity in DCs[23-25]. In candidiasis, IDO activity was induced at sites of infection as well as in DCs via IFN-γ- and cytotoxic T lymphocyte-associate antigen (CTLA) 4-dependent mechanisms. IDO inhibition greatly exacerbated the infection and associated inflammatory pathology, as a result of deregulated innate and adaptive immune responses. In vitro, IDO blockade reduced IL-10 production in response to hyphae and increased IL-6/IL-12 production in response to yeasts by PP-DCs. Consistent with the finding that PP-DCs producing IL-10 are absolutely required for activation of CD4+ CD25+ Treg capable of negatively regulating the inflammatory response and antifungal Th1 immunity upon adoptive transfer in vivo (see below)[26], the number of IL-10-producing CD4+ CD25+ Treg was significantly decreased in mice with candidiasis upon IDO blockade. Concomitantly, the number of CD4+ T cells producing IFN-γ increased, while that of cells producing IL-4 would decreased[27]. It appears that activation of IL-10-producing CD4+ CD25+ T cells is one important mechanism through which the IFN-γ/IDO-dependent pathway may control the local inflammatory pathology and Th1 reactivity to the fungus. These results provide novel mechanistic insights into complex events that, occurring at the fungus/pathogen interface, relate to the dynamics of host adaptation to the fungus. The production of IFN-γ may be squarely placed at this interface, where IDO activation likely exerts a fine control over inflammatory and adaptive antifungal responses. Therefore, the selective expression of IDO in the gut may represent the missing tissue-dependent factor that conditions the ability of DCs to produce IL-10 upon exposure to *Candida* hyphae, ultimately dictating the local pattern of both cytokine production and Th reactivity to the fungus. In addition, as *C. albicans* is a commensal of the human gastrointestinal and genitourinary tracts and IFN-γ is an important mediator of protective immunity to the fungus, the IFN-γ/IDO axis may accommodate fungal persistence in a host environment rich in IFN-γ. In its ability to downregulate antifungal Th1 response in the gastrointestinal tract, IDO behaves in a fashion similar to that described in mice with colitis, where IDO expression correlates with the occurrence of local tolerogenic responses[28].

7. DAMPENING INFLAMMATION AND ALLERGY TO FUNGI THROUGH Treg

It is clear that Treg cells, in addition or instead of Th2 cells or cells infected or exposed to pathogen products, are a principal source of antiinflammatory IL-10. Beyond their fundamental role in ensuring self-tolerance, different types of Treg cells participate in all immune responses[29]. Naturally occurring Treg cells originate in the thymus during the normal process of maturation and survive in the periphery as natural regulators, whereas inducible or adaptive Treg cells develop from conventional CD4[+] T cells that are activated in conditions of blockade of costimulatory signals, and presence of deactivating cytokines or drugs. CD4[+]CD25[+] Treg cells have several modes of suppressive action at their disposal, ranging from the inhibitory cytokines IL-10 and TGF-β to cell-to-cell contact via the inhibitory CTLA-4 molecule. Both natural and inducible Treg cells have been described in infection, their activation occurring through antigen-specific and nonspecific mechanisms. Treg cells with immunosuppressive activity have been described in fungal infections[26,27,30,31]. Naturally occurring Treg cells operating in the respiratory mucosa may account for the lack of pathology associated with fungal clearance in *P. carinii*-infected mice. In mice with candidiasis, CD4[+]CD25[+] Treg cells producing IL-10 and TGFβ, by dampening Th1-sterilizing immunity, prevented sterilization of the fungus from the gastrointestinal tract and allowed fungal persistence and the occurrence of memory immunity. The induction of CD4[+]CD25[+] Treg cells in candidiasis was B7 costimulation-dependent and involved the IFN-γ/IDO dependent pathway (see above). Because both the recovery of the fungus from the gastrointestinal tract and the detection of underlying Th1 reactivity can fluctuate in healthy subjects, it is conceivable that Treg cells mediate tolerance to the fungus at the site of colonization. In this regard, it is intriguing that IL-6, known to have inhibitory activity on Treg-cell functionality, is not produced in response to the fungus at epithelial surfaces[32], a finding that may explain the downregulation of IFN-γ production in some patients with recurrent vaginal candidiasis[33]. Similarly, the defective Th1 cytokine production seen in association with high levels of IL-10 in patients with chronic mucocutaneous candidiasis[34] (CMC) could be a consequence of deregulated Treg cells. CMC, although encompassing a variety of clinical entities, has been associated with autoimmune polyendocrinopathy–candidiasis–ectodermal dystrophy, a condition in which the mutated gene has been shown to be involved in the ontogeny of CD25[+] Treg cells[35].

Distinct Treg populations capable of mediating antiinflammatory or tolerogenic effects are coordinately induced after exposure to *Aspergillus* conidia[31]. *A. fumigatus* is a termotolerant saprophyte that is associated with a wide spectrum of diseases in humans of infectious, inflammatory, and allergic nature. The inherent resistance to diseases caused by the fungus suggests the occurrence of regulatory mechanisms that provide the host with protection from infection and tolerance to allergy. Although playing an essential role in the initiation and execution of the acute inflammatory response to fungus and subsequent resolution of infection, PMNs may act as double-edged swords, as the excessive release of oxidants and proteases may be responsible for injury to organs and fungal sepsis. In allergy, respiratory tolerance is mediated by lung plasmacytoid dendritic cells (pDCs) producing IL-10, which induce the development of CD4[+] Treg cells, expressing membrane-bound TGF-β and Foxp3, in a costimulation and TLR-dependent fashion[36]. It has been demonstrated that a division of labor oc-

curs between functionally distinct Treg cells that are coordinately activated by a
CD28/B7-dependent costimulatory pathway after exposure of mice to *Aspergillus* resting conidia. Early in infection, inflammation is controlled by the expansion, activation, and local recruitment of CD4⁺CD25⁺ Treg cells expressing the
same phenotype as those that control intestinal inflammation and autoimmunity
and suppressing PMNs through the combined actions of IL-10 and CTLA-4 on
IDO. Late in infection, and similarly in allergy, tolerogenic Treg cells of the
same phenotype as those controlling graft-versus-host diseases or diabetes,
which produce IL-10 and TGF-β, inhibit Th2 cells and prevent allergy to the
fungus. Taken together, these observations suggest that the capacity of Treg
cells to inhibit aspects of innate and adaptive immunity may be central to their
regulatory activity in fungal infections. This may result in the occurrence of
immune responses vigorous enough to provide adequate host defense, without
necessarily eliminating the pathogen (which could limit immune memory) or
causing an unacceptable level of host damage (Figure 1).

8. LOOKING FORWARD

The therapeutic efficacy of antifungals is limited without the help of host immune reactivity. Various cytokines, including chemokines and growth factors,
have proved to be beneficial in experimental and human refractory fungal infections[37]. The Th1–Th2 balance itself can be the target of immunotherapy. The
inhibition of Th2 cytokines, or the addition of Th1 cytokines, can increase the
efficacy of antifungals, such as polyens and azoles, in experimental mycoses. In
the past decade, a dramatic shift has occurred in our mechanistic understanding
of innate immunity. Precisely, the appreciation that activation of the innate immune system initiates, amplifies, and drives antigen-specific immune responses
together with the identification of discrete cell types, specific receptors, and the
signaling pathways involved in the activation of innate immunity has provided a
multitude of new targets for exploitation by the developments of adjuvants for
vaccines. Developments in DC biology are providing opportunities for improved
strategies for the prevention and management of fungal diseases in immunocompromised patients. The model has brought dendritic cells to center stage as
promising targets for intervention for immunotherapy and vaccine development
and has shifted the emphasis from the "antigen" toward the "adjuvant." The ultimate challenge will be to design fungal vaccines capable of inducing optimal
immune responses by targeting specific receptors on DCs. This will require,
however, further studies aimed at elucidating the convergence and divergence of
pathways of immune protection elicited in infections or upon vaccination.

9. ACKNOWLEDGMENTS

We thank Lara Bellocchio for dedicated editorial assistance. This study was supported by the the National Research Project on AIDS, contract 50F.30, "Opportunistic Infections and Tuberculosis," Italy.

10. REFERENCES

1. L. Romani. Overview of fungal pathogens. In *Immunology of infectious diseases*, pp. 25–37. Ed. S.H. Kaufmann, A. Sher and R. Ahmed. Washington, DC: ASM Press (2001).
2. N.A. Gow, A.J. Brown and F.C. Odds. Fungal morphogenesis and host invasion. *Curr Opin Microbiol* **5**:366–371 (2002).
3. M.C. Lorenz and G.R. Fink. The glyoxylate cycle is required for fungal virulence. *Nature* **412**:83–86 (2001).
4. L. Romani, F. Bistoni and P. Puccetti. Fungi, dendritic cells and receptors: a host perspective of fungal virulence. *Trends Microbiol* **10**:508–514 (2002).
5. R.A. Calderone. *Candida and candidiasis*. Washington, DC: ASM Press (2002).
6. K.A. Marr, T. Patterson and D. Denning. Aspergillosis: pathogenesis, clinical manifestations, and therapy. *Infect Dis Clin North Am* **16**:875–894 (2002).
7. L. Romani. Immunity to fungal infections. *Nat Rev Immunol* **4**:1–23 (2004).
8. S. Akira and K. Takeda. Toll-like receptor signalling. *Nat Rev Immunol* **4**:499–511 (2004).
9. L. Romani. The T cell response against fungal infections. *Curr Opin Immunol* **9**:484–490 (1997).
10. L. Romani. Innate immunity to fungi: the art of speed and specificity. In *Pathogenic fungi: host interactions and emerging strategies for control*, pp. 167–214. Ed. G. San-Blas and R.A. Calderone. Norfolk: Caister Academic Press (2004).
11. S.L. Newman. *Histoplasma capsulatum*: diary of an intracellular survivor. In *Fungal pathogenesis, principles and clinical applications*, pp. 81–96. Ed. R.A. Calderone and R.L. Cihlar. New York: Marcel Dekker (2002).
12. L.A. Gildea, R.E. Morris and S.L. Newman. Histoplasma capsulatum yeasts are phagocytosed via very late antigen-5, killed, and processed for antigen presentation by human dendritic cells. *J Immunol* **166**:1049–1056 (2001).
13. S. Akira, K. Takeda and T. Kaisho. Toll-like receptors: critical proteins linking innate and acquired immunity. *Nat Immunol* **2**:675–680 (2001).
14. S. Bellocchio, C. Montagnoli, S. Bozza, R. Gaziano, G. Rossi, S.S. Mambula, A. Vecchi, A. Mantovani, S.M. Levitz and L. Romani. The contribution of the Toll-like/IL-1 receptor superfamily to innate and adaptive immunity to fungal pathogens in vivo. *J Immunol* **172**:3059–3069 (2004).
15. S. Bellocchio, S. Moretti, K. Perruccio, F. Fallarino, S. Bozza, C. Montagnoli, P. Mosci, G.B. Lipford, L. Pitzurra and L. Romani. TLRs govern neutrophil activity in aspergillosis. *J Immunol* **173**:7406–7415 (2004).
16. M.G. Netea, J.W. Van der Meer and B.J. Kullberg. Toll-like receptors as an escape mechanism from the host defense. *Trends Microbiol* **12**:484–488 (2004).
17. S.M. Levitz. Interactions of Toll-like receptors with fungi. *Microbes Infect* **6**:1351–1355 (2004).

18. K. Shortman and Y.J. Liu. Mouse and human dendritic cell subtypes. *Nat Rev Immunol* **2**:151–161 (2002).

19. A. Bacci, C. Montagnoli, K. Perruccio, S. Bozza, R. Gaziano, L. Pitzurra, A. Velardi, C.F. d'Ostiani, J.E. Cutler and L. Romani. Dendritic cells pulsed with fungal RNA induce protective immunity to Candida albicans in hematopoietic transplantation. *J Immunol* **168**:2904–2913 (2002).

20. S. Bozza, R. Gaziano, A. Spreca, A. Bacci, C. Montagnoli, P. di Francesco and L. Romani. Dendritic cells transport conidia and hyphae of Aspergillus fumigatus from the airways to the draining lymph nodes and initiate disparate Th responses to the fungus. *J Immunol* **168**:1362–1371 (2002).

21. S. Bozza, K. Perruccio, C. Montagnoli, R. Gaziano, S. Bellocchio, E. Burchielli, G. Nkwanyuo, L. Pitzurra, A. Velardi and L. Romani. A dendritic cell vaccine against invasive aspergillosis in allogeneic hematopoietic transplantation. *Blood* **102**:3807–3814 (2003).

22. U. Grohmann, F. Fallarino and P. Puccetti. Tolerance, DCs and tryptophan: much ado about IDO. *Trends Immunol* **24**:242–248 (2003).

23. U. Grohmann, C. Orabona, F. Fallarino, C. Vacca, F. Calcinaro, A. Falorni, P. Candeloro, M.L. Belladonna, R. Bianchi, M.C. Fioretti and P. Puccetti. CTLA-4-Ig regulates tryptophan catabolism in vivo. *Nat Immunol* **3**:1097–1101 (2002).

24. F. Fallarino, U. Grohmann, K.W. Hwang, C. Orabona, C. Vacca, R. Bianchi, M.L. Belladonna, M.C. Fioretti, M.L. Alegre and P. Puccetti. Modulation of tryptophan catabolism by regulatory T cells. *Nat Immunol* **4**:1206–1212 (2003).

25. C. Orabona, U. Grohmann, M.L. Belladonna, F. Fallarino, C. Vacca, R. Bianchi, S. Bozza, C. Volpi, B.L. Salomon, M.C. Fioretti, L. Romani and P.C. Puccetti. CD28 induces immunostimulatory signals in dendritic cells via CD80 and CD86. *Nat Immunol* **5**:1134–1142 (2004).

26. C. Montagnoli, A. Bacci, S. Bozza, R. Gaziano, P. Mosci, A.H. Sharpe and L. Romani. B7/CD28-dependent CD4+CD25+ regulatory T cells are essential components of the memory-protective immunity to Candida albicans. *J Immunol* **169**:6298–6308 (2002).

27. S. Bozza, F. Fallarino, L. Pitzurra, T. Zelante, C. Montagnoli, S. Bellocchio, P. Mosci, C. Vacca, P. Puccetti and L. Romani. A crucial role for tryptophan catabolism at the host/*Candida albicans* interface. *J Immunol* **174**:2910–2918 (2005).

28. G.J. Gurtner, R.D. Newberry, S.R. Schloemann, K.G. McDonald and W.F. Stenson. Inhibition of indoleamine 2,3-dioxygenase augments trinitrobenzene sulfonic acid colitis in mice. *Gastroenterology* **125**:1762–1773 (2003).

29. Y. Belkaid and B.T. Rouse. Natural regulatory T cells in infectious disease. *Nat Immunol* **6**:353–360 (2005).

30. S. Hori, T.L. Carvalho and J. Demengeot. CD25+CD4+ regulatory T cells suppress CD4+ T cell-mediated pulmonary hyperinflammation driven by *Pneumocystis carinii* in immunodeficient mice. *Eur J Immunol* **32**:1282–1291 (2002).

31. C. Montagnoli, F. Fallarino, R. Gaziano, S. Bozza, S. Bellocchio, T. Zelante, W.P. Kurup, L. Pitzurra, P. Puccetti and L. Romani. The plasticity of dendritic cells at the host/fungal interface. *J Immunol* **204**:582–589 (2001).

32. P.L. Fidel Jr. The protective immune response against vaginal candidiasis: lessons learned from clinical studies and animal models. *Int Rev Immunol* **21**:515–548 (2002).

33. L.P. Carvalho, O. Bacellar, N. Neves, A.R. de Jesus and E.M. Carvalho. Downregulation of IFN-gamma production in patients with recurrent vaginal candidiasis. *J Allergy Clin Immunol* **109**:102–105 (2002).

34. D. Lilic. New perspectives on the immunology of chronic mucocutaneous candidiasis. *Curr Opin Infect Dis* **15**:143–147 (2002).

35. A. Liston, S. Lesage, J. Wilson, L. Peltonen and C.C. Goodnow. Aire regulates negative selection of organ-specific T cells. *Nat Immunol* **4**:350–354 (2003).

36. M. Ostroukhova, C. Seguin-Devaux, T.B. Oriss, B. Dixon-McCarthy, L. Yang, B.T. Ameredes, T.E. Corcoran and A. Ray. Tolerance induced by inhaled antigen involves CD4(+) T cells expressing membrane-bound TGF-beta and FOXP3. *J Clin Invest* **114**:28–38 (2004).

37. L. Romani. Host immune reactivity and antifungal chemotherapy: the power of being together. *J Chemother* **13**:347–353 (2001).

Author Index

Amouzegar, Taba K., 113
Bellocchio, Silvia, 209
Belz, Gabrielle T., 31
Benencia, Fabian, 185
Bonifazi, Pierluigi, 209
Bozza, Silvia, 209
Buckanovich, Ron, 185
Carlson, Louise M., 1
Carragher, Damian M., 55
Cejas, Pedro J., 1
Chappell, Craig P., 139
Conejo-Garcia, Jose R., 185
Coukos, George, 185
Corthay, Alexandre, 195
Della Chiesa, Mariella, 89
Dolfi, Douglas V., 149
Dondero, Alessandra, 89
Fallarino, Francesca, 209
Ferranti, Bruna, 89
Gaziano, Roberta, 209
Hayakawa, Yoshihiro, 103
Jacob, Joshy, 139
Katsikis, Peter D., 149
Kupresanin, Fiona, 31
Lee, Kelvin P., 1
Li, Jian-Ming, 69
Lindner, Inna, 1
Lund, Frances E., 171
Marcenaro, Emanuela, 89

Montagnoli, Claudia, 209
Mount, Adele M., 31
Moretta, Alessandro, 89
Moretti, Silvia, 209
Moyron-Quiroz, Juan, 55
Partida-Sánchez, Santiago, 171
Plano, Gregory V., 1
Puccetti, Paolo, 209
Pulendran, Bali, 43
Querec, Troy D., 43
Randall, Troy D., 55
Rangel-Moreno, Javier, 55
Rivero-Nava, Laura, 171
Roberts, Paul C., 113
Romani, Luigina, 209
Shi, Guixiu, 171
Smith, Christopher M., 31
Smyth, Mark J., 103
Swanborg, Robert H., 113
Swann, Jeremy, 103
ten Berge, Ineke J.M., 121
Torruellas, Julie, 1
Trivedi, Prachi P., 113
van Leeuwen, Ester M.M., 121
van Lier, René A.W., 121
Waller, Edmund K., 69
Wilson, Nicholas S., 31
Wolf, Norbert A., 113
Zelante, Teresa, 209

Subject Index

A

Activating signals, 90
Acute myelogenous leukemia (AML) and
   dendritic cell differentiation, 7
Adaptive immunity
     to fungi, 211, 214–215
     and innate immune responses, 92–93
     regulation of, 113–118
Adenosine diphosphate ribose (ADPR), 172,
   179
Adoptive transfer, 141, 145–147
     of tumor-specific CD4⁺ T cells,
       196–197
AFC, 146–147
Ag
     and memory B cell transfer, 145–147
     specificity, 139–140
Allogenic hematopoietic progenitor cell
   transplantation (HPCT), 69–84
Angiogenesis
     pathological, 185
     physiological, 185
     tumor, 187
Anthrax and dendritic cell differentiation, 15
Antifungal immunity, 209–218
Antigenic rechallenge, 160–163
Antigen presentation
     capacity of, 11
     and dendritic cells, 31–39
Antigen-presenting cells (APCs), 55, 82
     and CD8+ T cells, 155–156
     and donor T cell alloreactivity, 70–84
     as endothelial cells, 186–187
Antigens
     persistence and memory CD8+ T cells,
       126–127
     in pulmonary immune response, 56
     and regulation of IL-7Rα, 130
     stimulation and dendritic cells, 2–3
     and T cell responses, 155–160
Antitumor activity, 106, 190–191, 195
     and natural killer cells, 92

Aspergillus, 210, 214–215
Aspergillus conidia, 217–218
Aspergillus fumigatus, 210
Autoimmune diseases and natural killer
   cells, 114

B

B7, 150, 154
BB-1, 153
4-1BB, 158
     stimulation, 161
B cells
     activation of, 55
     and β-galactosidase expression, 142–
       147
     germinal center-derived, 139–147
     in the lungs, 60
     responses to nasal associated lymphoid
       tissue (NALT), 58
Bone marrow
     CD11b depletion, 73
     and graft-versus-leukemia, 77–79, 83
     memory T cells, 118
Bone marrow-derived cells and antigen
   presentation, 34
Bone marrow-derived natural killer cells,
   117–118
Bone marrow grafts, 73
     transplanting, 75–77
Bone marrow transplantation, 71, 186
     and memory T cells content, 79–80
Bronchus-associated lymphoid tissue
   (BALT), 60–63

C

Calcium mobilization, 176–177
Cancer. See also Tumors
     immunosurveillance, 197–204
     immunotherapy, 196–197
     and T cells, 195–205
Candida, 210, 214–215, 216

Candida albicans, 210
CCL2, 173
CCL19, 57, 58, 173
CCL21, 56–57, 58, 62, 173
CCR2, 174
CCR7, 127–128
CCR7⁻CD62L⁺, 127–128
CCR7⁻CD62L⁻, 127–128
CD3 and costimulation of T cells, 158
CD3+ thymocytes, 153
CD4, 32
CD4⁺CD25⁺ Treg cells, 217
CD4 cells and dendritic cells, 2
CD4 memory T cells, 74
CD4⁺ T cells, 36, 60
    activating macrophages, 201–204
    activation of in lymph nodes, 199–200
    and adoptive transfer in cancer immu-
      notherapy, 196–197
    in cancer immunosurveillance, 197–204
    and costimulation, 150, 151
    and memory CD8+ T cells, 130, 161
    and persistent viruses, 123
    responses to yellow fever vaccine (YF-
      17D), 46
CD8α, 32
CD8αDC, 36, 37–38
CD8+ cells
    and CD27, 162–163
    and CD28, 161–162
    and graft-versus-leukemia, 83
CD8+ memory T cells, 74
    function of, 126–127
    generation of, 124–125
CD8⁺ T cells, 58–59, 60
    amplification of, 38
    and antigen-presenting cells (APCs),
      155–160
    and cancer, 196, 197
    and CD4+ T cells, 161
    and cell responses to pathogens, 31–39
    and costimulation, 150, 151
    maintenance of, 129–130
    and persistent viruses, 122–131
    phenotype of, 127–129
    priming of, 33–34, 36
    response to pathogens, 35
    response to yellow fever vaccine (YF-
      17D), 45
CD11b, 32, 70–71, 117
    in bone marrow graft, 79–80
    depletion from bone marrow, 73
    and graft-versus-leukemia, 78–79, 83

CD11b⁻CD8DC, 36
CD11b+ cells, 82
    and graft-versus-host disease, 75–77
CD11b⁻DC, 36
CD11b^low^DC, 37
CD14+, 186
CD16+ natural killer cells, 91
CD27, 128
    costimulation of CD8+ T cells, 162–163
    costimulation of T cells, 155, 158–160
CD28, 128
    and costimulation of T cells, 150,
      153–154
    and memory responses, 161–162
CD31, 179, 189–190
CD34, 186, 189–190
CD38
    and chemokine receptors, 178–179
    and dendritic cell migration, 173–174
    and lung inflammatory responses,
      172–173
    and neutrophil migration, 173
    regulating chemokine receptor signal-
      ing, 176–178
    regulating T cell immune responses,
      174–176
CD40, 2, 7
CD45+/CD11c+, 187–189
CD45RA⁻, 128
CD56^bright^, 117
CD56+ natural killer cells, 91
CD62L, 127–128, 130
CD69, 199
CD70, 159
CD133+, 186
CD146, 189–190
CD154, 7
CD205, 32
CD38KO mice, 172–173
Cell cycle inhibition by natural killer cells,
    116–117
Cell sorting, 141
Cell-to-cell contact, 90, 115
Chemokine receptor signaling, 176–178
Chemokines, 56, 58
    in lymphoid organs, 62–63
Chronic mucocutaneous candidiasis (CMC),
    217
Chronic myelogenous leukemia (CML) and
    dendritic cell differentiation, 7
Classical activation pathway, 202–203
Collagen gel, 197–198
Constitutive defense mechanism, 211

Cooption, 186
Costimulation, 149–150
    and T cell development, 151–155
CR3, 212–214
CTLA4, 14, 218
CTL responses to viruses, 158
CXCL8, 94
CXCL12, 173
CXCL13, 56, 58, 62
Cyclic adenosine diphosphate ribose
    (cADPR), 172, 176–177
Cyclin D3, 116
Cyclosporine A, 156
Cytokines
    in antifungal immunity, 211
    and antitumor activities, 196
    in bone marrow transplantation, 81
    and dendritic cell differentiation, 6–7
    and maintenance of memory CD8+ T
        cells, 129
    and mast cells, 95
    and memory responses, 160–163
    and natural killer cells, 90–91, 92–93
    and NKG2D activation, 106
    production by eosinophils, 95–96
    and tumor-specific CD4$^+$ T cells,
        199–200
Cytomegalovirus (CMV), 122, 123–124,
    127
    and IL-7Rα regulation, 130–131
Cytotoxicity
    mediated by natural killer cells, 118
    receptors, 113

D

DC-SIGN, 94
β-defensin, 188–189
Dendritic cells (DCs)
    and antifungal immunity, 214–215
    and bone marrow and spleen grafts, 73
    and bone marrow transplantation, 81–84
    and CD38, 173–174
    dermal, 32
    differentiation and signal transduction,
        1–17
    differentiation and Yersinia, 16–17
    disruption of differentiation, 15–17
    donor-derived and transplantation,
        70–84
    as endothelial cells, 186–187
    functional diversity of, 1–5
    generation of, 4–5

    and innate immunity of yellow fever
        vaccine 17D, 46–48
    interstitial, 32
    lineage, 5
    maturation of, 2
    and natural killer cells, 114
    in pathogen responses, 33–38
    precursors and bone marrow transplan-
        tation, 81–84
    progenitors, 1–6
    resident, 32
    of spleen and lymph nodes, 31–33
    subsets of, 31–33, 34–38
    and tryptophan catabolism, 215–216
Dengue virus, 46–47
Diacylglycerol (DAG), 7–8
DN1, 154
DN3, 154
DN4, 154
DNA sequencing, 141
Double-negative (DN) dendritic cells, 32
DTDR-CD11c transgenic animals, 161

E

E2F, 116
Ecto-enzymes, 171
Effector cells, 55
    differentiation of, 161
    generation of, 124–125
    regulated by mast cells, 95
ELISPOT assay, 141
Endothelial cells
    antigen-presenting cells as, 186–187
    progenitors, 186
Endothelization, 188–190
Eosinophils and innate immunity, 95–97
E protein in yellow fever virus, 44
Epstein-Barr virus (EBV), 122, 124, 127
ERK pathway, 10
Experimental autoimmune encephalomye-
    litis (EAE) and natural killer cells,
    114, 118
Extracellular stimuli and dendritic cell dif-
    ferentiation, 6–7

F

FcγRII/III, 202–203
FC portion of immunoglobulins (FcRs),
    211, 213
Fetal-maternal tolerance, 114, 118
Fibrosarcoma and NKG2D ligands, 105

Flaviviruses, 46–47
Flt3 ligand (Flt3L), 6–7
Fungal morphogenesis, 209
Fungal pathogens, 210
Fungi
　　immunity to, 209–218
　　sensing, 211–214

G

β-galactosidase, 139, 140, 197
　　expression of, 142–144
　　and memory response transfer, 145–147
Gamma interferon. *See* Interferon-gamma
Germinal center-cre transgenic mice, 141–142
Germinal centers, 59–60
　　in B cells, 139–147
GM-CSF, 6
G-protein coupled receptors (GPCRs), 176
Graft rejection, 75
Graft-versus-host disease (GvHD), 70, 74, 75–77, 82
　　and graft-versus-leukemia, 77–79
Graft-versus-leukemia (GvL), 77–79, 83
Gut infection and dendritic cell subsets, 37–38

H

Hematopoiesis, 6
Hematopoietic cells in neoangiogenesis, 186
Hematopoietic engraftment, 74
Hematopoietic stem cells (HSCs), 152
Hepatitis B virus (HBV), 122
Hepatitis C virus (HCV), 122, 126, 128
Herpes virus, 123
High endothelial venules (HEVs), 55
Human immunodeficiency virus (HIV), 122, 126–127, 128
　　and IL-7Rα regulation, 130
Hypermutation of β-galactosidase, 145–147

I

iDC (immature), 1–2
IgG, 59
IgG1, 156
IgG2a, 156
IgG2b, 156
IgG antibodies, 45
IgM, 59
　　antibodies, 45

IκB kinase (IKKβ), 16
IL-1 receptors, 213
IL-2 and natural killer cells, 116
IL-4, 2, 96
　　in antifungal immunity, 211
　　and natural killer cells, 92, 95
IL-7, 129
IL-7Rα, 127, 129
　　regulation of, 130–131
IL-10, 2, 94, 216, 218
IL-12, 2, 90, 96
　　and natural killer cells, 92
IL-15, 129
IL-18, 91
　　and NKG2D, 106
IL-21 and NKG2D, 106
Immune homeostasis, 118
Immune polarization, 70
Immune system
　　and persistent viruses, 123–124
　　responses regulated by CD38, 174–176
Immunity
　　and dendritic cells, 35–38
　　post-transplant, 70–84
　　pulmonary, 57–63
Immunization and β-galactosidase expression, 142–144
Immunoglobulins and CD28 T cells, 156
Immunoregulation and natural killer (NK) cells, 90–92
Immunostimulation, 2
Indoleamine 2,3-dioxygenase (IDO), 216
Inducible bronchus associated lymphoid tissue (iBALT), 60–63
Inducible defense mechanism, 211
Infections
　　and dendritic cell subsets, 35–38
　　and persistent viruses, 122
Inflammation
　　dampening of, 216–218
　　and lymphoid tissue formation, 62
　　response and natural killer cells, 89–92
Influenza, 128
　　and immunity, 57, 58–59, 62–63
Innate immunity
　　and adaptive immune responses, 92–93
　　and eosinophils, 94–97
　　to fungi, 210–211
　　and mast cells, 94–97
　　and natural killer cells, 89–97
　　and neutrophils, 95–97
　　and tumor growth, 103–107
　　of yellow fever vaccine 17D, 46–48

Inteferon-α/β, 104
Interferon-γ, 2, 159, 216
    in antifungal immunity, 211
    in bone marrow transplantation, 74, 81
    and memory CD8+ T cells, 126
    and T cell-mediated macrophage
        activation, 202–203
Interferon-producing cells (IPC), 104
Intracellular signal transduction, 7–15
Invariant natural killer cells (iNKT), 5
Ionomycin, 114, 116

J

c-Jun pathway, 10

K

KG1 cells, 11
KG1a cells, 11–12
Kinases and dendritic cell differentiation,
    9–10

L

λ_1 V regions, 144–145
Langerhans cells, 32, 36
LBRM tumor cell line, 72, 79, 83
Lck and costimulation, 151
Leukemia
    and bone marrow transplantation, 77–79
    and dendritic cell differentiation, 7
Leukocyte migration, 177
Listeria monocytogenes, 35, 159, 162
LTα, 62
LT$\alpha^{-/-}$ mice, 57–58
LTβR, 56
Lung
    immune responses of, 55–64
    infection and dendritic cell subsets,
        36–37
    inflammatory responses on CD38,
        172–173
Lymph nodes
    and CD4$^+$ T cells, 199–200
    and dendritic cells, 31–33
    structure and development, 55–57
Lymphocytes, costimulation of, 149–150,
    152
Lymphocytic choriomeningitis virus
    (LCMV), 35, 122, 125, 126, 157
    and CTL responses to, 158
Lymphoid neogenesis, 62

Lymphoid organs in pulmonary immunity,
    57–63
Lymphoid Tissue inducer cells (LTi cells),
    56
Lymphoid tissues in pulmonary immune
    responses, 55–64
Lymphotoxin (LT) signaling pathway,
    56–57

M

Mac-1, 94
Macrophages
    activation by CD4$^+$ T cells, 201–204
    and injected myeloma cells, 200–201
Major histocompatibility complex (MHC),
    186
Major histocompatibility complex class I
    (MHC-I)
    and natural killer cells, 113–114
    presentation, 35
Major histocompatibility complex class Ib
    (MHC-Ib), 104–106
Major histocompatibility complex class II
    (MHC-II), 195
    and injected myeloma cells, 200–201,
        204–205
    presentation, 36
Mannosyl receptors, 211–212
MAPK/ERK pathway, 9
MAP kinase and dendritic cell differentia-
    tion, 9–10
MAP kinase kinases (MKKs), 16
Mast cells and innate immunity, 94–97
Maternal-fetal tolerance, 114, 118
Matrigel, 197–198
Mediastinal lymph node (MLN), 55
Memory B cells, 139–147
    and adoptive transfer, 145–147
Memory CD8$^+$ T cells, 196
    maintenance of, 129–130
    phenotype of, 127–129
Memory responses, 160
    transfer of, 145–147
Memory T cells, 38, 118
    in bone marrow grafts, 79–80
    function of, 126–127
    generation of, 124–125
    and persistent viruses, 122–131
Methylcholanthrene (MCA)-induced
    tumors, 104, 105
MHC. *See* Major histocompatibility com-
    plex (MHC)

MICA, 104
MICB, 104
Mice models for persistent viruses, 122
Microbe sensing, 211–214
Migratory tissue-derived dendritic cells, 36–38
Monocyte-derived dendritic cells (MDDC)
  activation of, 93–94
  and natural killer cells, 89–92
Monocytes
  as endothelial cells, 186–187
  as hematopoietic cell progenitors, 186
MOPC315, 197, 200, 203
Multiple sclerosis and natural killer cells, 114
Murine cytomegalovirus (MCMV), 122, 125
Murine γ-herpesvirus 68, 122
Mutations in β-galactosidase B cells, 144–145
MyD88 pathway, 213, 215
Myelin basic protein (MBP), 114–115

N

NAD
  analogues, 177
  interaction with CD38, 179–180
Naive T cells, priming of, 33–34, 36
Nasal Associated Lymphoid Tissue (NALT)
  and antigens, 56
  structure and function of, 57–58
Natural killer (NK) cells, 89, 113
  bone marrow-derived, 117–118
  cell cycle inhibition, 116–117
  decidual, 114
  in immune homeostasis, 118
  immunoregulatory role of, 90–92
  and NKG2D, 106
  peripheral, 114
  regulatory function of, 114–115
  splenic, 117
  and T cells, 115–116
N(8-Br-A)D+, 177
Neoangiogenesis, 186
Neovascularization, 185
Neutralizing antibody responses, 45
Neutrophils
  in adaptive immunity, 93–94
    migration and CD38, 173
NFκBinhDC, 14
NFκB inhibition, 3–4

NFκB pathway, 8, 9, 10
  and dendritic cell differentiation, 10–15
  inhibition of during dendritic cell differentiation, 13–15
Nicotinamide adenine dinucleotide (NAD(P)), 172
Nicotinic acid adenine dinucleotide (NAADP), 172, 179
NK1.1^{bright}, 117
NK+CD3⁻ cells, 114–115
NKG2D, 104–106, 113
  activation by cytokines, 106
NKp30, 91, 113
NKp44, 113
NKp46, 113
Notch Delta ligands, 152
Notch 1 receptor, 152
Notch 2 receptor, 152
Notch receptors, 152

O

Opsonins, 215
Ovalbumin (OVA) and CD38 regulation, 174–175
Ovarian carcinoma, 187–188, 191

P

p21, 117, 118
Pathogen-associated molecular patterns (PAMP), 90, 92
Pathogens and dendritic cells, 31–39
Pathological angiogenesis, 185
Pattern recognition receptors (PRRs), 90, 211
PCR, 141
pDC, 35
Perforin, 115, 117
Peripheral lymph node addressin, 62
Peroxisome proliferative activated receptor-γ (PPARγ), 5
Persistent viruses. See Viruses, persistent
Peyer's patches and antigens, 56
Phagocytic cells, 33–34
Phagocytosis, 214–215
Phorbol esters, 7–8
Phorbol-12-myristate-13 acetate (PMA), 114, 116
  and protein kinase C activation, 8, 9
Phospholipids in signal transduction, 7–8

Physiological angiogenesis, 185
PKCβII, 9, 12
Plasmacytoid dendritic cells (PDC), 32
    and natural killer cells, 89–92
PMA. *See* Phorbol-12-myristate-13 acetate
    (PMA)
PNA⁺ B cells, 143–145
PNAd, 62
Post-transplant immunity, 70–84
Post-transplant lymphoproliferative disease
    (PTLD), 124
Post-transplant relapse, 70, 82
p38 pathway, 10
p50 protein, 10, 11
p52 protein, 10, 11
p100 protein, 11
Protein kinase C (PKC), and dendritic cell
    differentiation, 7–9
Pulmonary immune responses, 55–64

R

RACK proteins, 8
Rae-1 and tumorigenesis, 105
RAG2⁻/⁻ hosts, 146
Regulatory T cells, 211, 216–218
RelA protein, 10
RelB protein, 10, 11–13
c-Rel protein, 10
Respiratory syncytial virus (RSV), 128

S

S. pneumoniae, 172–173
Sarcomas and NKG2D ligands, 105
Sca1+, 186
Secondary infections, 38
Self/non-self discrimination, 150
Sentinel cells, 214
Signaling pathways, and dendritic cell dif-
    ferentiation, 6–15
Signalling and T cell development, 151–155
Signal transduction and dendritic cell differ-
    entiation, 1–17
Signal transduction pathways, disruption of,
    15–17
SLP mice, 58–59
Spleen
    and antigens, 56
    and β-galactosidase expression,
        142–143
    and dendritic cells, 31–33
    grafts, 73

Splenic T cell purification, 73
Splenocytes
    and bone marrow transplantation, 75–77
    and graft-versus-host disease, 75–77
    and graft-versus-leukemia, 77–79
    and memory T cells, 79–80
STAT5, 116
Stem cell content in bone marrow graft, 75
Stem cell factor (SCF), 7

T

T cell receptor. *See* TCR
T cells
    activation, 1–4, 55, 71, 82
    antigen specific responses, 155–160
    in bone marrow grafts, 79–80
    and cancer, 195–205
    costimulation of, 150
    and dendritic cell differentiation, 13–14
    development of, 151–155
    inhibition by natural killer cells,
        116–117
    inhibition of proliferation, 114–116
    priming, 33–34
    proliferation by bone marrow-derived
        natural killer cells, 117–118
    purification of in spleen, 73
    regulation by CD38, 174–176
    responses to CD11b, 71
    responses to nasal associated lymphoid
        tissue (NALT), 58
    responses to yellow fever vaccine (YF-
        17D), 45
    tolerance, 71
TCR, 197
    stimulation and IL-7Rα, 130
    and T cell development, 151–154
    transgenic cell, 35
Th1, 94, 217, 218
    polarization, 70
    response, 2
    and tumor elimination, 196–197
Th1/Th2 balance, 47
Th2, 218
    polarization, 70
    response, 2
    and tumor elimination, 196
Thymic development, 153–155
Tip DC, 35–36
TLR2, 94, 214
TLR3, 90, 95
TLR4, 94, 214

TLR9, 90, 91–92

TNF-alpha, 14, 94–95

TNF and costimulation, 150

Toll-like receptors (TLRs), 3, 211, 213–214
    activation by yellow fever vaccine (YF-17D), 43, 46, 47
    and natural killer cells, 90, 92
    and neutrophils, 94

Treg, 211, 216–218

Tryptophan catabolism, 215–216

Tumor angiogenesis, 187

Tumor-associated macrophages (TAMs), 203–204

Tumor cells, and T cell-activated macrophages, 203–204

Tumorigenesis
    and innate immunity, 103–107
    and NKG2D ligands, 105–106

Tumor necrosis factor-alpha (TNF-α). See TNF-alpha

Tumors. See also Cancer
    and CD8⁺ cells, 196, 197
    cells, 197–198
    innate immunity to, 107
    and vascular leukocytes, 190–191

Type I interferon and tumorigenesis, 103–104

U

3'-UTR, 44

V

Vascular endothelial growth factor (VEGF), 187, 188–190

Vascular leukocytes (VLCs), 187–190
    and antitumor immune response, 190–191

Vasculogenesis, 185

VE-Cadherin, 189–190

VEGFR1, 189–190

VEGFR2, 189–190

Vesicular stomatitis virus (VSV), 157

Viral clearance, 63

Viremia, 44–45

Viruses
    and CTL responses to, 158
    and dendritic cell differentiation, 15
    and dendritic cell subsets, 36
    persistent
    and CD8+ T cells, 124–131
    effects on host immune system, 123–124
    prevalence of, 122

W

West Nile virus, 46–47

Y

Yellow fever vaccine 17D and immunity (YF-17D), 43–48

Yersinia
    and dendritic cell differentiation, 16–17
    and disruption of intracellular signaling pathways, 18

YopJ, 15, 16–17

YopP, 17

Yops, 16